YOUR GUIDE TO THE
2024
SOLAR
ECLIPSE

Michael E. Bakich

Kalmbach
Media

Kalmbach Media
21027 Crossroads Circle
Waukesha, Wisconsin 53186
www.myscienceshop.com

Published in 2022
27 26 25 24 23   2 3 4 5 6

Manufactured in China

ISBN: 978-1-62700-913-3
EISBN: 978-1-62700-914-0

Editor: Eric White
Book Design: Lisa Schroeder

Library of Congress Control Number: 2022933217

CONTENTS

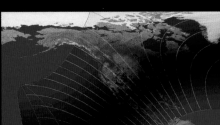

INTRODUCTION

The original meaning of ἔκλειψη (the Greek word for "eclipse") is a forsaking, quitting, or disappearance. So, the covering of one object by another describes what an eclipse is. Total solar eclipses are exact lineups of the Sun, the Moon, and Earth, in that order.

Our solar system is a group of a few large and many small objects. The main one is the Sun. Imagine a line between the sun and any other object. Because everything in the solar system is in motion, that line will point in a different direction as time passes. The line also shows the direction of the object's shadow, which will be opposite the Sun.

Every so often, a third object aligns with the other two. If the two that aren't the Sun are close enough, the shadow from the closest one to the Sun may fall on the other. It may completely cover the second object or only partially cover it. Likewise, the first object may completely or partially block out the Sun's disk. When this occurs, eclipses happen.

The larger a body is, the longer its shadow. At Earth's average distance from the Sun, Earth's umbral (dark, inner) shadow has an average length of 855,000 miles and the Moon's is 255,000 miles long. These numbers vary a little because the distances of these bodies from the Sun change. Still, it's easy to see why total lunar eclipses last much longer than total solar eclipses. The disk of Earth's shadow is much larger than the corresponding disk of the Moon's shadow at the average Earth-Moon distance of about 238,900 miles.

This photographer captured the stages of the total solar eclipse on August 21, 2017. The exposures from left to right roughly match the position and motion of the eclipse during totality. Alan Dyer

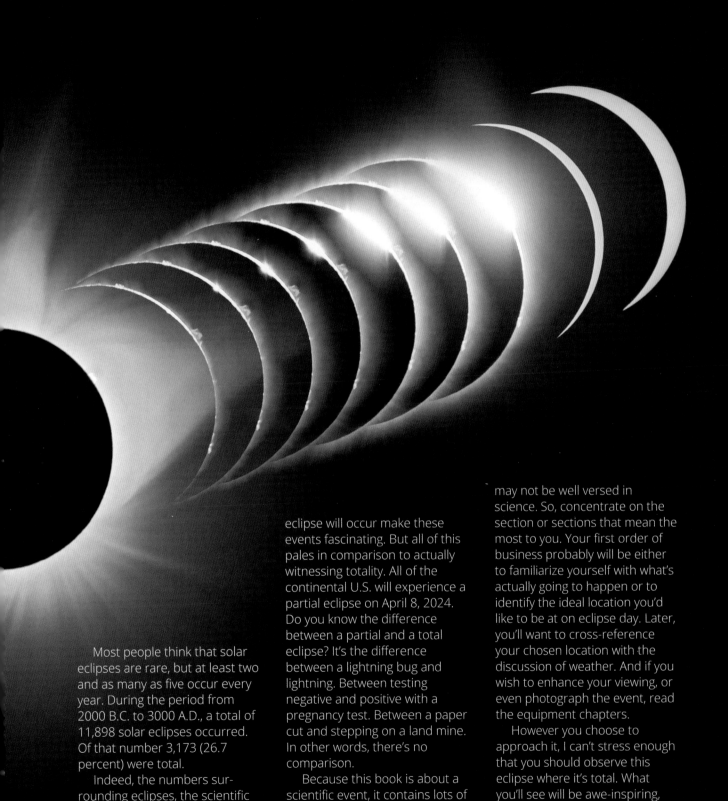

Most people think that solar eclipses are rare, but at least two and as many as five occur every year. During the period from 2000 B.C. to 3000 A.D., a total of 11,898 solar eclipses occurred. Of that number 3,173 (26.7 percent) were total.

Indeed, the numbers surrounding eclipses, the scientific reasons they happen, and the way astronomers can predict—to a fraction of a second—where, when, and for how long a given eclipse will occur make these events fascinating. But all of this pales in comparison to actually witnessing totality. All of the continental U.S. will experience a partial eclipse on April 8, 2024. Do you know the difference between a partial and a total eclipse? It's the difference between a lightning bug and lightning. Between testing negative and positive with a pregnancy test. Between a paper cut and stepping on a land mine. In other words, there's no comparison.

Because this book is about a scientific event, it contains lots of facts. But it's also meant to appeal to astronomy newbies, people who certainly will be interested in this event, but who may not be well versed in science. So, concentrate on the section or sections that mean the most to you. Your first order of business probably will be either to familiarize yourself with what's actually going to happen or to identify the ideal location you'd like to be at on eclipse day. Later, you'll want to cross-reference your chosen location with the discussion of weather. And if you wish to enhance your viewing, or even photograph the event, read the equipment chapters.

However you choose to approach it, I can't stress enough that you should observe this eclipse where it's total. What you'll see will be awe-inspiring, and you'll never forget it. It will stand out as one of the great-est—if not the greatest—sights you'll ever see.

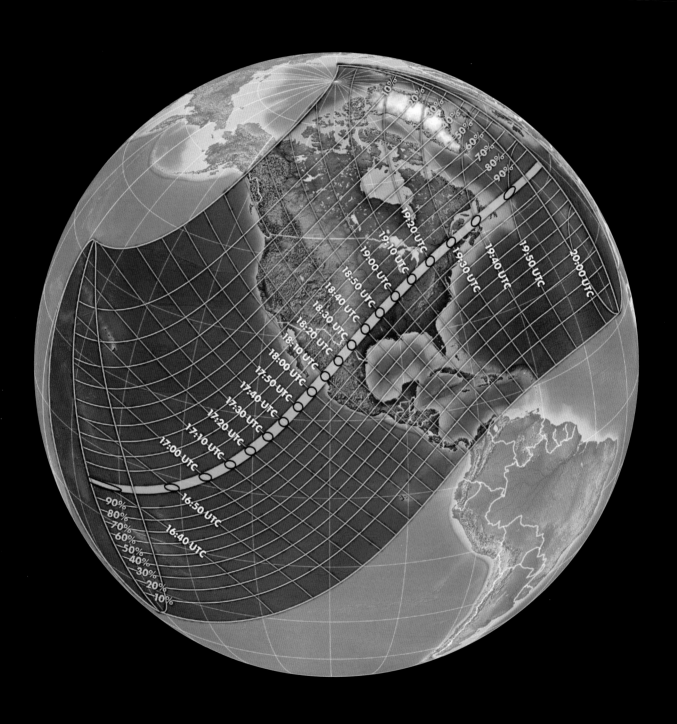

The path of the April 8, 2024, total solar eclipse stretches from the Pacific to the Atlantic Ocean. To convert UTC to Eastern Daylight Time, just subtract 4 hours. For CDT, subtract 5 hours, and so on.

1

WHAT'S THE BIG DEAL?

Drama is coming to North America. On April 8, 2024, Sun-watchers along a thin curved line that stretches for thousands of miles through just three countries will experience nature's grandest spectacle: a total solar eclipse.

Astronomy clubs, government agencies, cities— even whole states—are preparing for large numbers of visitors seeking to experience darkness at midday.

The Moon's dark inner shadow, which is the only place the eclipse will be total, touches no other land on Earth except for Mexico, the U.S., and a tiny bit of Canada. The Moon's shadow speeds along thousands of miles of open ocean waters in the Pacific before it contacts Mexico, and then in the Atlantic after leaving Canada, so these three countries are the only places you can stand on solid ground to view the event.

This will be the first total solar eclipse crossing the continental U.S. in 7 years. Prior to 2017, there was a 38-year gap. One did cover Hawaii in 1991, but in the 48 contiguous states, the previous solar eclipse occurred on February 26, 1979. Unfortunately, not many people saw it because it was visible from just five states in the Northwest. Making matters worse, that winter's weather for the most part was bleak along the path of totality. Before the 1979 eclipse, you have to go back to March 7, 1970, when a total solar eclipse traveled up the East Coast of the U.S., again occurring in a scant five states.

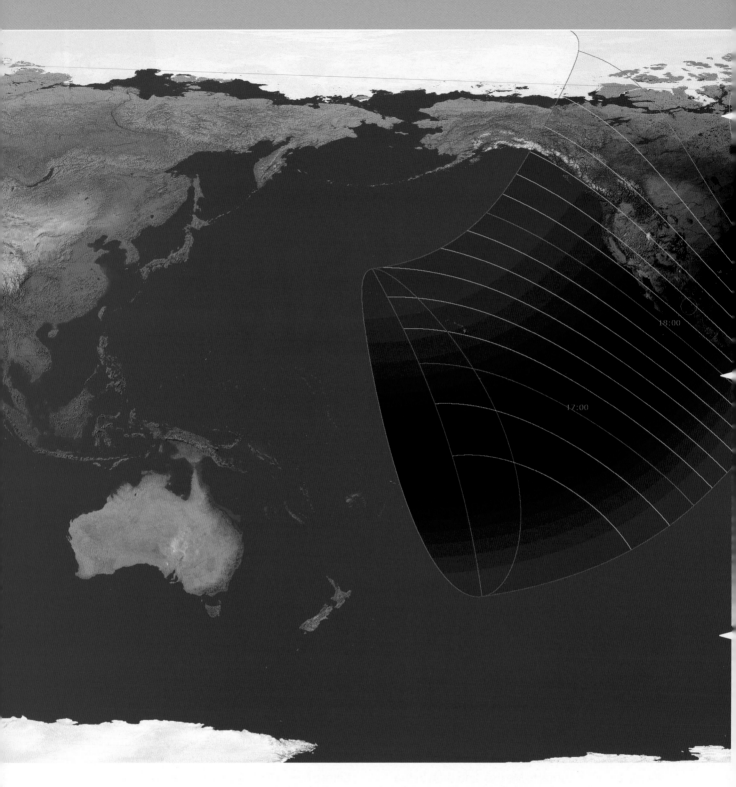

Although total solar eclipses occur more often than total lunar eclipses, more people have seen a total eclipse of the Moon. Few, on the other hand, have seen a total solar eclipse. The reason is simple. We live on Earth, and it's our perspective that interacts with the geometry of these events. During a lunar eclipse, anyone on the night side of our planet under a clear sky can see the Moon passing through Earth's dark inner shadow. That shadow, even as far away as the Moon, is quite a bit larger than the Moon, so it takes our satellite some time to pass through it. In fact, if the Moon passes through the center of Earth's shadow the total part of the eclipse can last as long as 106 minutes. Usually, totality doesn't last that long because the Moon passes either above or below the center of

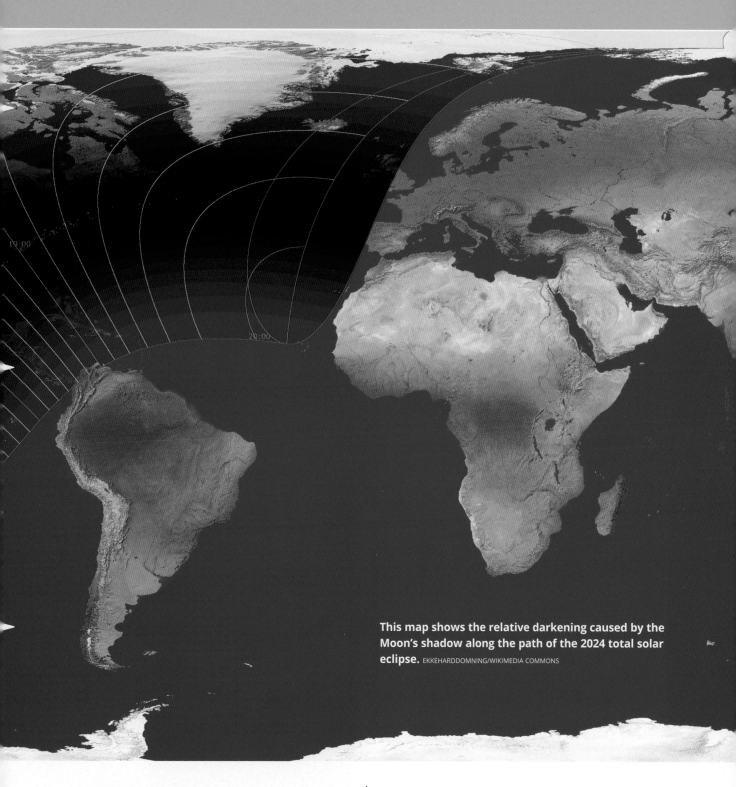

This map shows the relative darkening caused by the Moon's shadow along the path of the 2024 total solar eclipse. EKKEHARDDOMNING/WIKIMEDIA COMMONS

the shadow our planet casts.

Conversely, the Moon and its shadow at the distance from Earth are much smaller; so small, in fact, that the shadow barely reaches our planet's surface. Anybody in the lighter outer region of the shadow will see a partial solar eclipse. The lucky individuals under the dark inner shadow will experience a total solar eclipse. But not for long.

Solar totality lasts a maximum of 7½ minutes. In fact, the longest extent of totality in the 5,000-year span from 2000 B.C. to 3000 A.D. is 7 minutes and 29 seconds, which will be the eclipse that occurs on July 16, 2186. We won't be nearly that fortunate on April 8, 2024. The maximum duration of totality then will be 4 minutes and 28 seconds.

Now, I do want to say a few words about the event. It's all about totality. Everyone in the continental U.S. will see at least a partial eclipse. In fact, if you have clear skies on eclipse day, the Moon will cover at least 16 percent of the Sun's surface. And that's from the northwestern tip of Washington. Although the Moon covering part of the Sun's disk sounds cool, you need to aim higher.

Likening a partial eclipse to a total eclipse is like comparing almost dying to actually dying. If you're outside during a solar eclipse with 16 percent coverage, you won't even notice your surroundings getting darker. And it doesn't matter whether the partial eclipse above your location is 16, 56, or 96 percent. Only totality reveals the true celestial spectacles: the two diamond rings, the Sun's glorious corona, 360° of sunset colors, and stars in the daytime. So, remember, to see any of this you must be in the path of totality.

To capture this time-lapse photo of the 2019 total solar eclipse at El Molle, Chile, images were taken at 5-minute intervals using a Canon 5D Mk III DSLR mounted on a tripod with a Canon 24-70mm zoom lens set at 30mm and f/11.

The diamond-ring effect occurrs at the beginning and end of totality during a total solar eclipse. As the last bits of sunlight pass through the valleys on the Moon's limb, and the faint corona around the sun is just becoming visible, it looks like a ring with glittering diamonds on it. NASA/CARLA THOMAS

Knowing all of this, you want to try to be on the center line. The fact that the Moon's shadow is round probably isn't a revelation. If it were square, it wouldn't matter where you viewed totality. People across its width would experience the same duration of darkness. The shadow is round, however, so the longest eclipse occurs at its center line because that's where you'll experience the full width of the Moon's shadow.

Another important thing to keep in mind is that this event will happen! Some of the problems astronomers deal with are due to the uncertainty and limited visibility of some celestial events. Comets may appear bright if their compositions are just so. Meteor showers might reach storm levels if we pass through a thick part of the dust stream (and generally the best views occur after midnight). A supernova as bright as a whole galaxy may be visible, but you still may need a telescope to view it. In contrast to such events, this solar eclipse will occur at the exact time astronomers predict, along a precisely plotted path, and for the lengths of time given. Guaranteed. And it's a daytime event to boot!

The next total solar eclipse that causes the

This graphic illustrates why you should try to position yourself on the eclipse's center line, represented by the red arrow. If you're near the northern or southern limit, represented by the yellow arrow, the duration of totality will be much less. HOLLEY Y. BAKICH

Moon's shadow to fall across the continental U.S. won't happen for 20 years. It occurs on August 23, 2044. And, similar to the 1979 event, that one will be visible only in Montana and North Dakota. Total solar eclipses follow in 2045 and 2078, and they'll be great ones.

But it's 2024 that's causing all the excitement now. And you've got the right reference for it in your hands. The spectacular maps by Michael Zeiler will show you exactly where the eclipse is happening. Indeed, the main purpose of this book is to keep you informed so that you can approach this event without a shadow of a doubt.

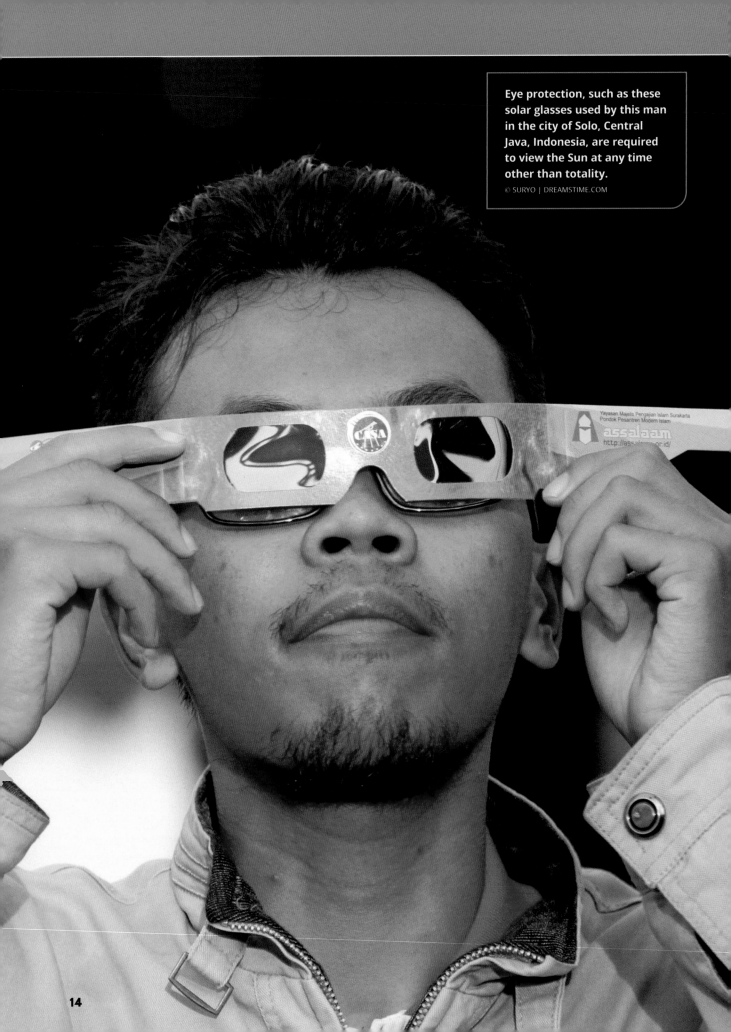

Eye protection, such as these solar glasses used by this man in the city of Solo, Central Java, Indonesia, are required to view the Sun at any time other than totality.
© SURYO | DREAMSTIME.COM

14

2

FREQUENTLY ASKED QUESTIONS

The purpose of this chapter is to answer some of the most important questions for both the general public and the media. Yes, the eclipse may be far off, but it's never too early for knowledge. Plus, these are the facts, and they won't change.

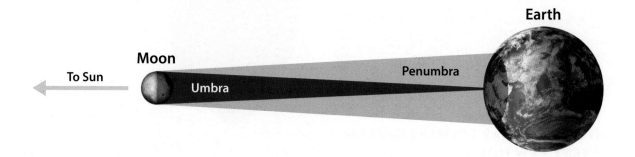

Earth

Moon

To Sun

Umbra

Penumbra

WHAT'S HAPPENING?

A solar eclipse is a lineup of the Sun, the Moon, and Earth—in that order. The Moon, directly between the Sun and Earth, casts a shadow on our planet. If you're in the dark part of that shadow, called the umbra, you'll see an eclipse. If you're in the light part, the penumbra, you'll see a partial eclipse.

If you're beneath the Moon's umbra, you'll experience a total solar eclipse. Anyone in the penumbra will only see a partial eclipse. HOLLEY Y. BAKICH

THE SUN IS A LOT BIGGER THAN THE MOON, SO HOW DOES THIS WORK?

While our daytime star's diameter is approximately 400 times larger than that of the Moon, it also lies roughly 400 times farther away. This means both disks appear to our eyes to be the same size.

WHEN DO SOLAR ECLIPSES OCCUR?

A solar eclipse happens at New Moon. The Moon has to be between the Sun and Earth for a solar eclipse to occur. The only lunar phase when that happens is New Moon.

WHY ARE SOME ECLIPSES LONGER THAN OTHERS?

The reason the total phases of solar eclipses vary in time is because Earth is not always at the same distance from the Sun, and the Moon is not always the same distance from Earth. The Earth-Sun distance varies by as much as 3 percent. That may not sound like much, but it's nearly 3 million miles. The Moon-Earth distance, meanwhile, can change by as much as 12 percent. The result is that the Moon's apparent diameter — that is, the disk that we see — can range from 7 percent larger than the Sun to 10 percent smaller than the Sun.

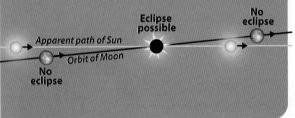

WHY DON'T SOLAR ECLIPSES HAPPEN AT EVERY NEW MOON?

The reason is that the Moon's orbit tilts 5° to Earth's orbit around the Sun. Astronomers call the two intersections of their orbits nodes. Solar eclipses only occur when the Sun lies at one node and the Moon is at its New phase. During most lunar months, the Sun lies either above or below one of the nodes, and no eclipse happens.

WHAT DO MAGNITUDE AND OBSCURATION MEAN?

The magnitude of a solar eclipse is the percent of the Sun's diameter that the Moon covers during the maximum eclipse. The obscuration is the percent of the Sun's total surface area covered at maximum. Here's an example: Let's say we just observed a partial eclipse where the Moon covered half the Sun's diameter. In this case, the magnitude of the eclipse equals 50 percent. However, the amount of obscuration (that is, the area of the Sun's disk that the Moon blotted out) was only 39.1 percent.

For a total solar eclipse, the obscuration always equals exactly 100 percent. The magnitude, however, can be anywhere from 100 percent, which astronomers would designate as 1.0000, to a bit more than 108 percent, or 1.0805, the magnitude of the total solar eclipse on July 16, 2186. Magnitudes greater than 100 percent simply mean that the Moon's apparent diameter is that much greater than the Sun's. For the April 8, 2024, eclipse, the magnitude will be 1.0566.

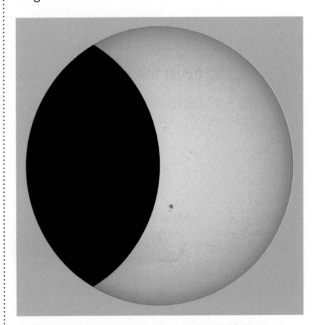

This image shows the Moon covering half the Sun's diameter. The magnitude of an eclipse at this moment would be 0.50, or 50 percent. The obscuration, however, that is, the percentage of the Sun's area covered, would be only 39.1 percent.

HOLLEY Y. BAKICH

WHAT DOES THE WORD "SAROS" MEAN?

This is the length of time between similar solar (and lunar) eclipses and represents their periodicity and recurrence, which repeat every 6,585.3 days. Two eclipses separated by one Saros cycle are similar. They occur with the Sun, the Moon, and Earth at the same relative positions. Also, the Moon's distance from Earth is nearly the same, and the eclipses happen at the same time of year. British astronomer Edmond Halley coined the term "saros" from an 11th-century Byzantine book.

WHAT'S THE PATH OF THE 2024 ECLIPSE?

Here are just a few facts. The Moon's umbra touches the coast of Mexico at 12:07 p.m. Mexican Pacific Daylight Time less than 12 miles southeast of Mazatlán. After that, the center line crosses through three Mexican states: Sinaloa, Durango, and Coahuila; 10 U.S. states: Texas, Oklahoma, Arkansas, Missouri, Illinois, Indiana, Ohio, New York, Vermont, and Maine; and four Canadian provinces: Ontario, Quebec, New Brunswick, and Newfoundland, and Labrador. The last point of contact with land is on the coast of Newfoundland, at the small town of Catalina, located on the eastern side of the Bonavista Peninsula.

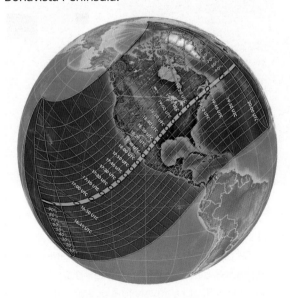

The path of the 2024 eclipse begins in the Pacific Ocean, crosses three countries, and ends in the North Atlantic Ocean. MICHAEL ZEILER

HOW LONG WILL TOTALITY BE?

On April 8, 2024, totality lasts a maximum of 4 minutes and 28 seconds. That's it. To experience that length, you'll need to be at Nazas, Mexico at 1:15 p.m. Mexican Central Daylight Time. This town of some 4,000 residents will surely see that number swell because it is the nearest location to the point of longest totality. The actual spot of longest totality is about 3 miles to the north, just east of Durango Paso Nacional, the road that connects Nazas to San Luis del Cordero.

Any device with small holes can project multiple images of the eclipsed Sun. HOLLEY Y. BAKICH

BESIDES TOTALITY, WHAT ELSE SHOULD YOU LOOK FOR?

Although the big payoff is the exact lineup of the Sun, the Moon, and your location, keep your eyes open during the partial phases that lead up to and follow it. As you view the beginning of the eclipse through a safe solar filter, around the three-quarters mark, you'll start to notice that shadows are getting sharper. The reason is that the Sun's disk is approaching a point, and a smaller light source produces better-defined shadows. At about 85 percent coverage, Venus will be visible 15° west-southwest of the Sun, and Jupiter will lie 30° east-northeast of the Sun. If any trees are at your viewing site, you may see their leaves act like pinhole cameras as hundreds of crescent Suns appear in their shadows.

HOW MANY PEOPLE WILL SEE THIS ECLIPSE?

Here's a question nobody can answer now, except in general terms. My opinion is that this eclipse will be one of the most viewed ever. Three factors will make it so: (1) the attention it will get from the media; (2) the superb coverage of the highway system in our country; and (3) the vast number of people who will have access to it from nearby large cities. The only possible negative is the typical weather throughout the U.S. on April 8, especially as the event tracks to the northeast.

WHAT EYE PROTECTION IS NEEDED DURING TOTALITY?

None. In fact, totality is safe to look at. During the time the Moon's disk covers that of the Sun, it's perfectly fine to look at the eclipse. In fact, to experience the awesomeness of the event, you must look at the Sun without a filter during totality.

WHAT ELSE WILL BE HAPPENING AS A RESULT OF THE ECLIPSE?

Nature will take heed. Depending on your surroundings, as totality nears you may experience strange things. Look: you'll notice a resemblance to the onset of night, though not exactly. Areas much lighter than the sky near the Sun lie all around the horizon. Shadows also look different. Listen: usually, any breeze will dissipate and birds (many of whom will come in to roost) will stop chirping. It is quiet. Feel: a 10°–15° Fahrenheit drop in temperature is not unusual.

To use solar viewing glasses, first inspect them for damage. Then, without looking at the Sun, put them on. Finally, find the Sun and enjoy. MICHAEL E. BAKICH

IS ANY EQUIPMENT REQUIRED?

You'll only need eclipse glasses or another type of approved solar filter to view the partial phases. A telescope is not needed. One of the great things about the total phase of a solar eclipse is that it looks best to naked eyes. The sight of the corona surrounding the Moon's black disk in a darkened sky is unforgettable. That said, binoculars give you a close-up view—but still at relatively low power—that you should take certainly advantage of during the event.

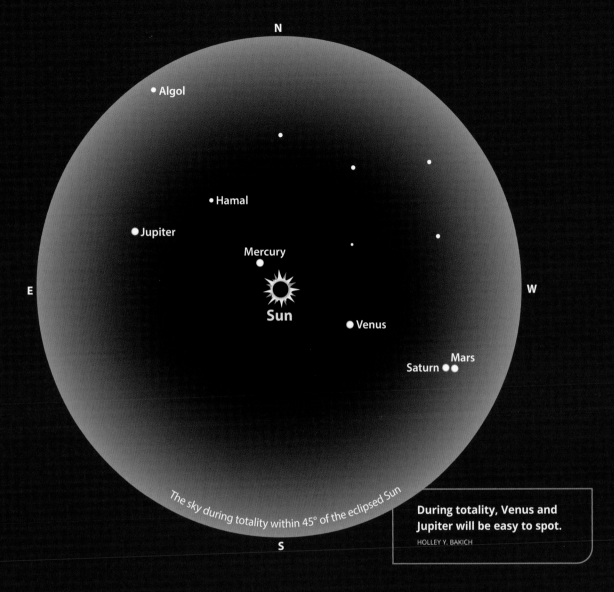

The sky during totality within 45° of the eclipsed Sun

During totality, Venus and Jupiter will be easy to spot.
HOLLEY Y. BAKICH

WHAT ARE THE DETAILS?

The Moon's lighter penumbra (under which anyone would experience a partial eclipse) first touches Earth at 11:42 a.m. EDT and last contacts it at 4:52 p.m. EDT. The total phase of the eclipse begins at 12:39 p.m. EDT, a bit more than 620 miles south of the Republic of Kiribati in the Pacific Ocean. The umbra remains in contact with Earth's surface for 3 hours, 16 minutes, and 45 seconds, until 3:55 p.m. EDT when it vanishes in the North Atlantic Ocean 340 miles southwest of Ireland. The total length of the eclipse path is 9,190 miles.

Greatest eclipse occurs at 2:17 p.m. EDT and the maximum length of totality anywhere on Earth is 4 minutes 28 seconds. That point is just a few miles north of the small town of Nazas, Mexico.

Venus will lie 15° west-southwest of the eclipsed Sun. Sighting this planet should be simple 15 minutes before totality. Because it's not all that far from the brilliant Sun, be cautious if you look for it through binoculars. Make absolutely certain you don't sweep the Sun into your field of view.

Jupiter will lie 30° east-northeast of the Sun. During the eclipse, Venus will blaze at magnitude –3.9, while Jupiter comes in second at magnitude –2.1. The difference, 1.8 magnitudes, means that Venus will shine 5¼ times as bright as Jupiter. Still, the king of the planets will certainly pop into view during totality, and perhaps a few minutes before.

Partial solar eclipse at sunrise from Southern England.
MARBO/DREAMSTIME.COM

GLOSSARY

So that we're all speaking the same language about solar eclipses, this chapter will briefly list of the terms you'll encounter, many of them with illustrations. Get familiar with them because you will see them again.

ALTITUDE—the height, in degrees, of a point or celestial object above the horizon. We measure altitude from 0° (on the horizon) to 90° (at the zenith, which is the overhead point). ASTRONOMY MAGAZINE/ROEN KELLY

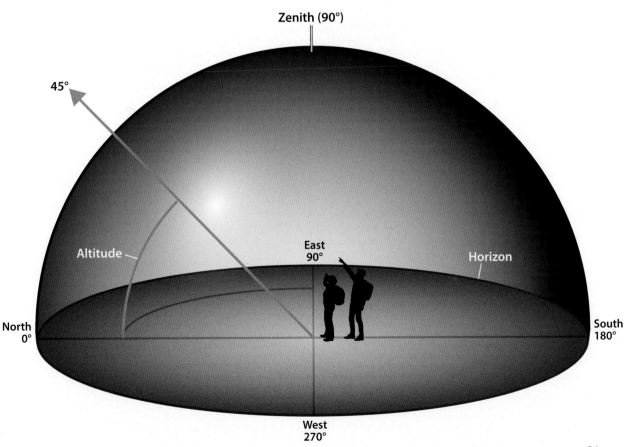

This is an angle of 1°. There are 360° in a circle. HOLLEY Y. BAKICH

ANGULAR DIAMETER—the apparent size of a celestial object, measured in degrees, minutes, and/or seconds, as seen from Earth. OK, let's define the three words in that sentence. A degree is 1/360 of a circle. Said another way, a circle contains 360°. A minute (short for a minute of arc or arc-minute) is 1/60 of 1°. A second (short for seconds of arc or arcsecond) is 1/60 of 1 minute of arc. So, 1° contains 3,600 arcseconds.

ANGULAR DISTANCE—this is the same thing as angular diameter except that we're measuring the distance between two objects, not the size of a single object. So the definition would be the distance between two celestial bodies expressed in degrees, minutes, and/or seconds of arc.

APHELION—the position of an object orbiting the Sun when it lies farthest from the Sun. Aphelion has two approved pronunciations: a FEEL ee on and ap HEEL ee on.

APOGEE—the position of the Moon or other object in Earth's orbit when it lies farthest from our planet.

AZIMUTH—the angular distance (from 0° to 360°) to an object measured eastward along the horizon starting from the north. So the azimuth of an object due north is 0°; due east would be 90°; south would be 180°; and a due-west azimuth equals 270°.

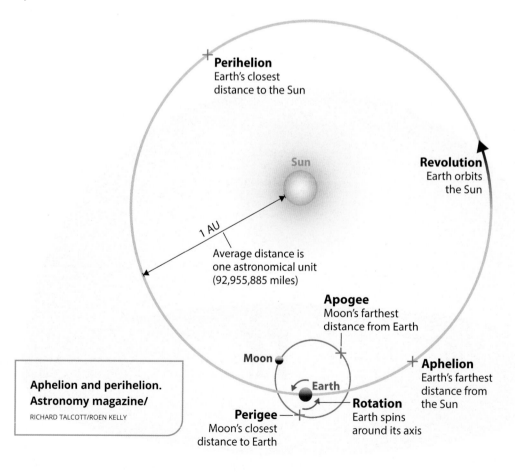

Perihelion
Earth's closest distance to the Sun

Sun

Revolution
Earth orbits the Sun

1 AU
Average distance is one astronomical unit (92,955,885 miles)

Apogee
Moon's farthest distance from Earth

Aphelion
Earth's farthest distance from the Sun

Moon

Earth

Rotation
Earth spins around its axis

Perigee
Moon's closest distance to Earth

Aphelion and perihelion.
Astronomy magazine/
RICHARD TALCOTT/ROEN KELLY

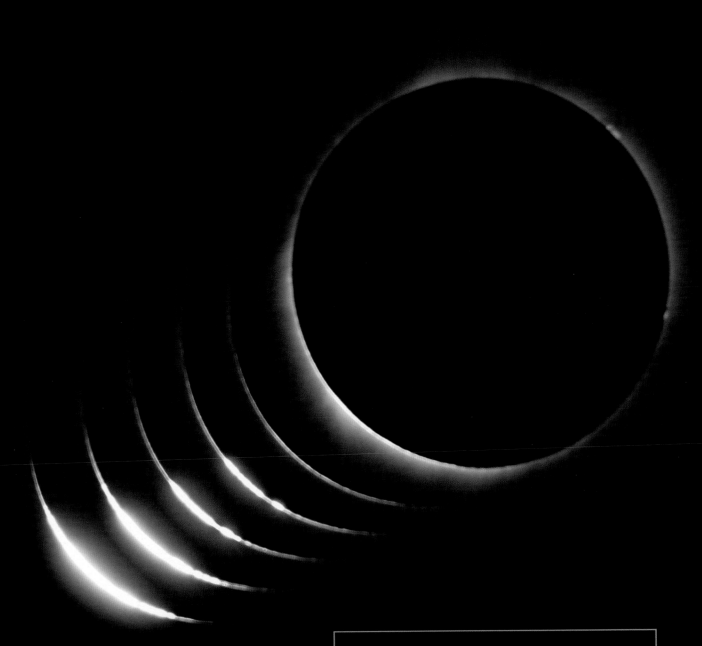

BAILY'S BEADS—during a total solar eclipse, the effect is often seen just before and just after totality when only a few points of sunlight are visible at the edge of the Moon. There, mountains block out the Sun's disk, but valleys permit light to shine through. Scientists named this phenomenon after English astronomer Francis Baily, who first explained the effect in 1836. RODNEY POMMIER

CENTER LINE—the midpoint of the width of the Moon's shadow on Earth; the center line is the location for the maximum duration of totality.

CHROMOSPHERE—the region of the Sun's atmosphere between its visible surface and its corona, sometimes briefly visible just before or after totality as an intense red glow at the Moon's edge.

CONJUNCTION—a point in the sky where two celestial bodies appear to line up. The lineup may be an exact one, as in the case of a total eclipse, or it may be a near one, as in the case of a New Moon (when our satellite is "in line" with the Sun).

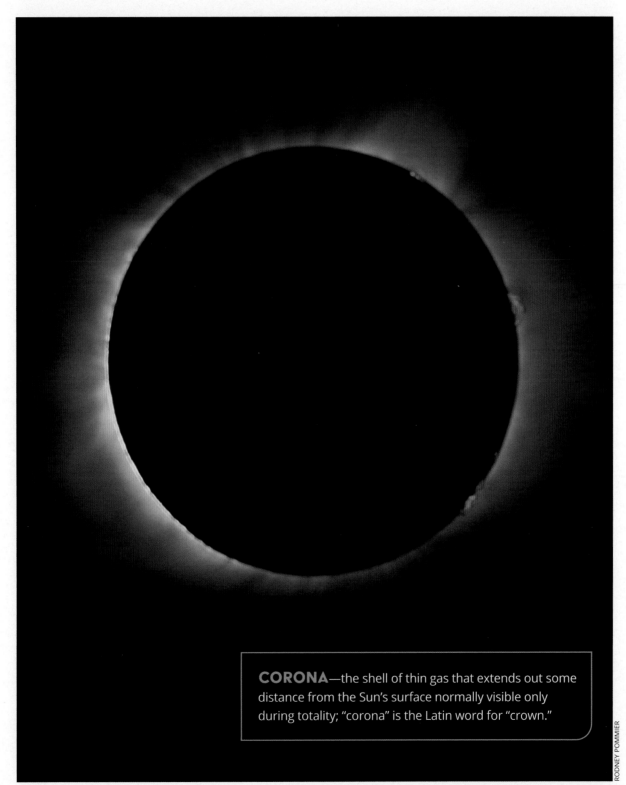

CORONA—the shell of thin gas that extends out some distance from the Sun's surface normally visible only during totality; "corona" is the Latin word for "crown."

RODNEY POMMIER

DIAMOND RING—the effect just before or just after totality during a total solar eclipse when a small portion of the Sun's disk plus its corona produce an effect similar to a ring with a brilliant diamond. RODNEY POMMIER

DISK—the visible surface of any heavenly body.

ECLIPTIC—the circle described by the Sun's apparent yearly path through the stars; the ecliptic traces the Sun's path through the constellations of the zodiac.

FLARE—a sudden burst of particles and energy from the Sun's photosphere; through a Hydrogen-alpha filter, flares often appear brighter than the surrounding area.

FOURTH CONTACT—during a solar eclipse, the moment that the disk of the Moon breaks contact with the Sun; this moment marks the end of the eclipse.

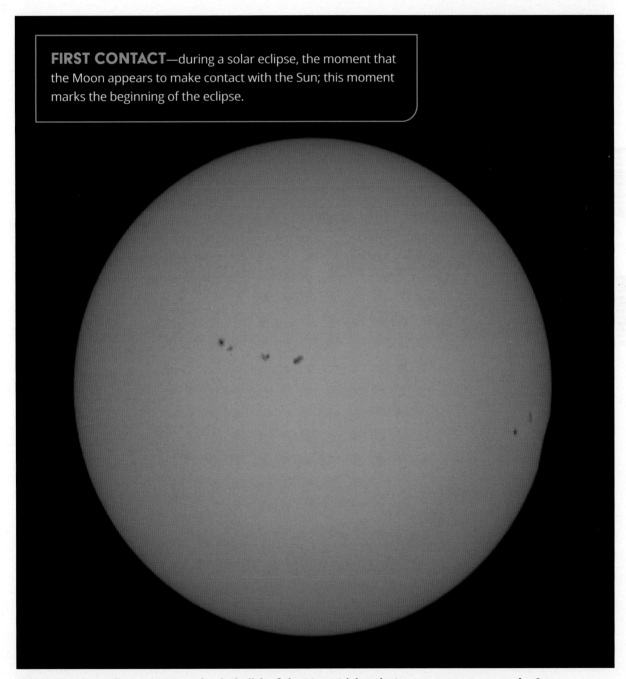

FIRST CONTACT—during a solar eclipse, the moment that the Moon appears to make contact with the Sun; this moment marks the beginning of the eclipse.

This image shows first contact as the dark disk of the Moon (right edge) starts to pass across the Sun.

RODNEY POMMIER

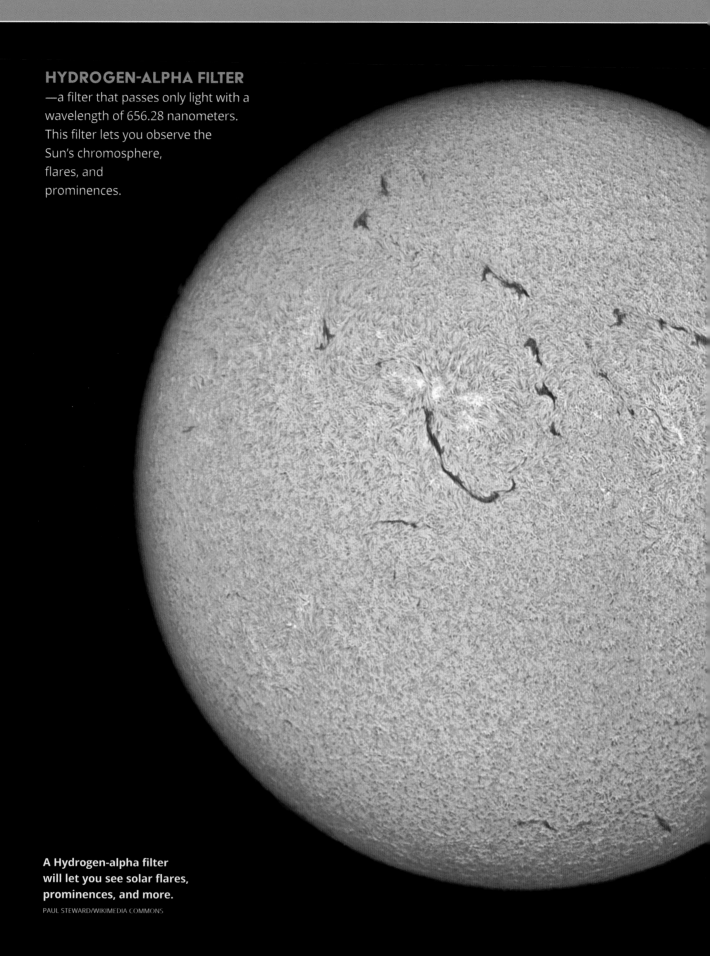

HYDROGEN-ALPHA FILTER
—a filter that passes only light with a wavelength of 656.28 nanometers. This filter lets you observe the Sun's chromosphere, flares, and prominences.

A Hydrogen-alpha filter will let you see solar flares, prominences, and more.

PAUL STEWARD/WIKIMEDIA COMMONS

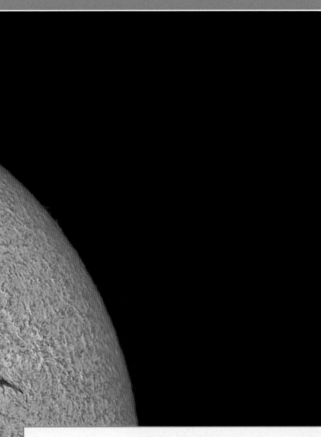

MAGNITUDE—the amount of the Sun's diameter the Moon covers during an eclipse; this is not the same as "obscuration."

NEW MOON—the phase where the Moon seems completely unlit from our perspective on Earth; the phase where the Moon is between Earth and the Sun; solar eclipses can occur only at New Moon.

NODES—the two places the plane of the Moon's orbit crosses the plane of Earth's orbit; eclipses can occur only near nodes.

OBSCURATION—the amount of the Sun's area the Moon covers during an eclipse; this is not the same as "magnitude."

ORBIT—the path of one celestial body around another.

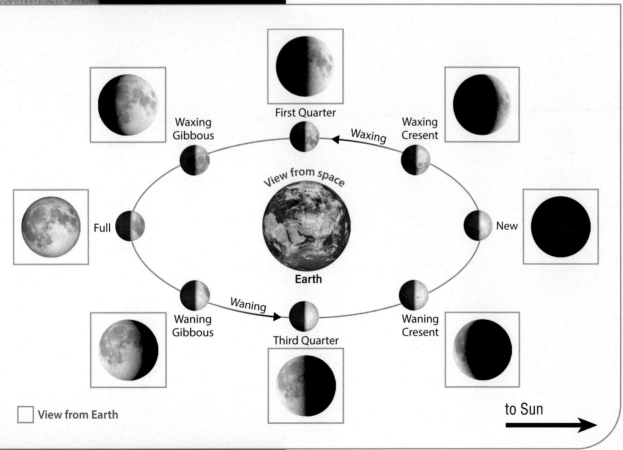

Lunar phases. Solar eclipses occur only at New Moon.
HOLLEY Y. BAKICH

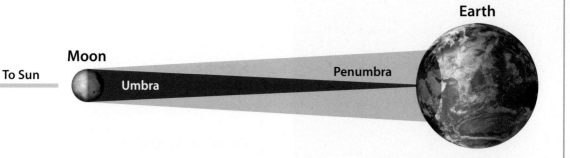

PENUMBRA—the less dark outer region of the Moon's shadow; an observer under the penumbra sees a partial solar eclipse. HOLLEY Y. BAKICH

PERIGEE—the position of the Moon or other object in Earth's orbit when it lies closest to our planet.

PERIHELION—the position of an object in solar orbit when it lies closest to the Sun.

PHOTOSPHERE—the visible surface of the Sun where our star emits visible light.

REVOLUTION—in astronomy, the orbiting of one body around another; Earth revolves around the Sun once each year.

ROTATION—the spinning of a celestial body on its axis; Earth rotates once a day.

SAROS CYCLE—a time period equal to 6,585.3 days between which similar eclipses occur.

SECOND CONTACT—during a total solar eclipse, the moment the Moon covers 100 percent of the Sun's disk; the instant totality begins.

SHADOW BANDS—faint ripples of light are occasionally seen on light-colored surfaces just before and just after totality.

PAUL STEWART/WIKIMEDIA COMMONS

PROMINENCE—a large-scale, gaseous formation above the surface of the Sun; during totality observers often see prominences at the Moon's dark edge.

SOLAR TELESCOPE—a telescope whose design lets you safely observe the Sun.

SUNSPOT—a temporarily cooler (and therefore darker) region on the Sun's disk caused by magnetic field variations.

SYZYGY—the lineup of three celestial bodies; for a solar eclipse, the lineup is the Sun, the Moon, and Earth.

UMBRA—the dark inner region of the Moon's shadow; anyone under the Moon's umbra will experience a total solar eclipse.

A Hydrogen-alpha telescope. CORONADO INSTRUMENTS

THIRD CONTACT—during a total solar
eclipse, this is the instant when totality ends.

It took a great deal of planning to get this spectacular photo of the 2019 total solar eclipse from the Atacama Desert in Chile. From finding the right location to experience totality, to ensuring you have the right equipment, it's never too soon to start preparing for eclipse day. GETTYIMAGES

A PLAN FOR ECLIPSE DAY

The solar eclipse will be a life-defining event that you won't want to have diminished with small annoyances. Follow the following common-sense tips, and you'll be ready.

1 • TAKE ECLIPSE DAY OFF—NOW!

You may think a year is a long lead time to get ready, but planning always pays off. Many potential eclipse watchers probably have the same idea. Figure out the earliest date that makes sense for you to request April 8, 2024, as a vacation day, and mark it on your calendar.

2 • MAKE A WEEKEND OUT OF IT

Eclipse day is on a Monday. Lots of related activities in locations touched by the Moon's inner shadow will occur on the preceding Saturday and Sunday. Find out what they are, where they're being held, and which you want to attend.

3 • ATTEND AN EVENT

You'll enjoy the eclipse more if you hook up with like-minded people. If you don't see any special goings on a few months before April 8, call your local astronomy club, planetarium, or science center. Anyone you talk to is sure to know of eclipse activities. If worrying about details isn't your thing, travel agents also offer trips that will allow you to experience the eclipse.

4 • GET INVOLVED

If your interests include public service, consider volunteering with a group putting on an eclipse event. You'll learn a lot and make some new friends in the process.

5 • FOCUS ON THE WEATHER

Meteorologists study chaotic systems. Nobody will know with absolute certainty what the weather will be like for a specific location on eclipse day, on April 8, 2024. Start watching the local weather for your chosen location on April 3.

Pay attention to the local weather starting April 3, 2024. KMBC-TV

A cruise ship full of excited eclipse chasers is a great way to enjoy the event. HOLLEY Y. BAKICH

6 • STAY FLEXIBLE ON ECLIPSE DAY

Unless you are certain that April 8 will be clear, don't do anything that would be hard to undo in a short time. For example, let's say you're taking a motor home to a certain city. You connect it to power, hook up the sewage hose, extend the awnings, set up chairs, start the grill, and more. But if it's cloudy six hours, three hours, or even one hour before the eclipse starts, you're going to want to move to a different location. Think of the time you would have saved if you had waited to set up. Also, the earlier you make your decision to move, the better. Traffic on eclipse day will be a nightmare.

7 • STAY FOCUSED

Totality will be the shortest three or four minutes of your life. All your attention should be on the Sun. Anything else is a waste. Be considerate of those around you. Please, no music.

If you're setting up a lot of equipment before the eclipse, be sure the weather won't force you to move quickly. MISSVAIN/WIKIMEDIA COMMONS

8 • PEE BEFORE THINGS GET GOING

This tip, above and beyond any other on this list, could be the most important one for you. Don't wait until 10 minutes before totality to start searching for a bathroom. Too much is happening then. Make a preemptive strike 45 minutes prior.

9 • RECORD THE TEMPERATURE

A point-and-shoot camera that takes movies will let you record the temperature as it drops. Here's a suggestion: Point your camera at a digital thermometer and a watch, both of which you previously attached to a white piece of cardboard or foam core board. Start recording video 15 or so minutes before totality and keep shooting until 15 minutes afterward. The results may surprise you.

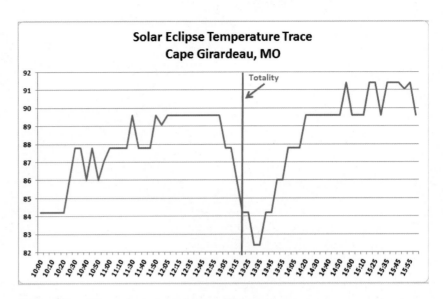

During the eclipse, the temperature will drop. NWS

10 • WATCH FOR THE MOON'S SHADOW

If your viewing location is at a high elevation, or even at the top of a good-sized hill, you may see the Moon's shadow approaching you. Seeing this isn't easy because as the shadow crosses Indianapolis, for example, it is moving at nearly 2,000 mph.

11 • VIEW THE 360° SUNSET

During totality, take just a few seconds to tear your eyes away from the sky and scan the horizon. You'll see sunset colors all around you. This happens because, in effect, those locations are where sunset or sunrise are happening.

12 • GET A FILTER IN ADVANCE

Cardboard eclipse glasses with lenses of safe optical Mylar cost about $2. Such a device—it's not a toy—will let you safely look directly at the Sun. Buy one well in advance, and you can look at the Sun anytime. Be sure to purchase it from a reputable vendor. Some vendors also carry filters for binoculars and telescopes. Another safe solar filter is a no. 14 welder's glass, but the Sun will look green through it.

A 360-degree twilight occurred during the eclipse at Orin Junction, Wyoming in 2017. This gave the illusion of the Sun setting below the horizon in every direction.
O'DEA/WIKIMEDIA COMMONS

A no. 14 welder's glass is an inexpensive way to safely observe the Sun. HOLLEY Y. BAKICH

13 • NO FILTER? YOU CAN STILL WATCH

Except during totality, you should never look at the Sun. But what if you've forgotten a filter? You can still watch by making a pinhole camera. This can be as simple as two pieces of paper with a tiny hole in one of them. Make the hole as round as you can with a pin or a sharp pencil. Hold the piece of paper with the hole above the other paper. The pinhole will allow light through to produce a tiny image of the sun that you can safely view. Moving the two pieces of paper farther apart will enlarge the Sun's image, but will also lessen its brightness. Work out a good compromise.

14 • BRING A CHAIR

In all likelihood, you'll be at your viewing site several hours before the eclipse starts. You don't really want to stand that whole time, do you?

15 • DON'T FORGET THE SUNSCREEN

Most people who go outside during daylight know this. Remember, even though it's still early spring, you'll be standing around or sitting outside for hours. You may even want to bring an umbrella for some welcome shade.

16 • TAKE LOTS OF PICTURES

Before and after totality, be sure to record your viewing site and the people with whom you shared this great event with. Speaking of sharing—a Sun-spotter solar telescope (sold as "STARLAB" by Walmart) provides a sharp image of the Sun that many people can view simultaneously.

17 • THE TIME WILL ZOOM BY

A totality that lasts nearly 4½ minutes might seem like a long time, but it zooms by faster than you can imagine. This is especially true if you're trying to photograph the Sun's ghostly corona because you have to remember so many things. No matter how much you've practiced (you have practiced, right?), the adrenaline rush you'll feel when the

Moon's shadow overtakes you and the cheers and screams you'll hear from those around you will certainly test your plan. So, please, remember my one cardinal rule for all eclipse photographers: If anything goes wrong, stop shooting and look!

18 • BRING SNACKS AND DRINKS

You're probably going to get hungry waiting for the eclipse to start. Unless you set up next to a convenience store, consider bringing something to eat and drink. And keep in mind that April is cool. A thermos with warm drinks is a great idea.

19 • BE PREPARED TO BE THE EXPERT

If you're planning an event or a family gathering related to the eclipse, consider this: Most likely, none of the people you encounter will have experienced darkness at noon. You will be the teacher for those attending. Equipment and filters will help your guests get more from the eclipse, but your own knowledge is the main tool that will help amplify their experience.

You can download instructions to make a Sun funnel at eclipse2017.nasa.gov/make-sun-funnel. HOLLEY Y. BAKICH

20 • INVITE SOMEONE WITH A SOLAR TELESCOPE

In the event you're thinking of hosting a private get-together, try to find someone with a telescope that has a solar filter to attend the event. While it's true that you don't need a scope to view the eclipse, having one there will generate quite a bit of buzz. And you (or the telescope's owner) can point out and describe sunspots, irregularities along the Moon's edge, and more.

If you don't have a solar filter, a pinhole in an object held in front of a white card will let you observe the Sun safely.
HOLLEY Y. BAKICH

21 • RESPECT TOTALITY

The eclipse of April 8, 2024, plus the events leading up to it will combine to be a fabulous social affair. Totality itself, however, is a time that you should mentally shed your surroundings and focus solely on the sublime celestial dance above you. You'll have plenty of time for conversations afterward. A get-together with family and friends after the eclipse will help you unwind a bit and hear what others experienced during the eclipse. However, a period for quiet observation and appreciation, without distractions should be the rule during totality.

22 • SCHEDULE AN AFTER-ECLIPSE PARTY OR MEAL

Once the eclipse winds down, you'll be on an emotional high for hours, and so will everyone else. There's no better time to get together with family and friends and chat about what you've just experienced together.

23 • RECORD YOUR MEMORIES

Sometime shortly after the eclipse, when the event is still fresh in your mind, take time to write, voice-record, or make a video of your memories, thoughts, and impressions. A decade from now, such a chronicle will help you relive this fantastic event. Have friends join in, too. Stick a video camera in their faces and capture 30 seconds from each of them. You'll smile each time you watch it.

Be patient after the eclipse, and you won't have to endure hours of traffic. HOLLEY Y. BAKICH

24 • DON'T BE IN A RUSH AFTERWARD

Traffic gridlock will be horrendous after the eclipse. And the sooner you leave after totality ends, the worse it will be. Relax. Let the parts of the eclipse between the third and fourth contacts play out. Many people will view this as "what we saw before totality, but in reverse," but it's part of the process which can still be viewed through approved filters.

A colander might not be the first thing most people think of as what to bring to an eclipse, but holding something with a series of holes over a light-colored surface can reveal the crescent shape of the partially eclipsed Sun in a safe-to-view fashion. © RICHARD GUNION | DREAMSTIME.COM

WHAT TO BRING TO THE ECLIPSE

This chapter applies to all people who will be traveling to see the eclipse, but especially those who will not be part of an organized travel group. You may observe the eclipse alone, with friends or family, or at a public event.

I thought it would be good to provide a checklist of both common and unusual items that I think readers should bring to the eclipse. Such an inventory could get out of hand quickly, so I limited it to 25 entries.

And to those planning to do this alone—of the 16 total solar eclipses I've traveled to see, only the first two were lone-wolf adventures. The other 14 excursions all involved a travel agent. Of those, I would rate 13 of them an A and the other one a B–. That's a pretty impressive average and one that argues in favor of joining an organized tour for many reasons.

☐ **SUNSCREEN** When someone says, "solar safety," this is what I think of. So should you. Pick a brand that has a sun protection factor (SPF) of at least 50. And here's something to note: If your bottle of sunscreen is more than three years old, replace it. That's the standard shelf life for this product.

☐ **WATER** April 8 will be warm in the Southwest U.S. and cooler as the shadow tracks to the northeast. But you still need to hydrate. Even large events may run out of this vital fluid. If you're driving, don't leave home without a case of bottled water.

☐ **APPROVED SOLAR FILTER** Whether you use eclipse glasses, a homemade filter using solar Mylar, or a no. 14 welder's glass, you'll need these to view the partial phases. Also, if you plan to view the partial phases through binoculars or a telescope, you'll need approved solar filters for each of them.

IVY MAIN/WIKIMEDIA COMMONS

☐ **CAMERA** Whether or not you're photographing the eclipse, you'll want to document the day and the activities surrounding it.

☐ **TRIPOD** If you're bringing a camera, that is.

☐ **BINOCULARS** A great way to get a close-up view of the corona. Learn all about binoculars in Chapter 7.

☐ **YOUR ECLIPSE GUIDE** OK, this eclipse guide.

☐ **FOOD OR SNACKS** Certainly this isn't as critical as water; I mean, you're not going to starve. Having some healthy snacks or some pre-made sandwiches can help you avoid fast food and give you more options in culinary-challenged communities.

☐ **PROOF OF VACCINATION** I'm writing this more than two years ahead of the eclipse, so, hopefully, COVID-19 will be behind us. If not, some events may require this or a recent negative test.

☐ **MEDICINE** Be sure you have any prescriptions you need to take with you. And some pain medication also is a good idea. Sometimes too much Sun gives some people headaches.

☐ **TOILET PAPER** Let's see, millions of people on the road, rest stops few and far between. ... You fill in the details.

☐ **HAND SANITIZER** for the same reason as the last item.

☐ **CHAIRS** Bring at least one chair (fold-up varieties pack best) for each person in your party. You're not going to want to stand for (a minimum of) 2½ hours. The best chairs you can bring offer the choice of sitting upright or reclining.

☐ **EXTRA EYEGLASSES** You won't forget the ones on your face, but something may happen to that pair.

☐ **KID STUFF** I have no children, so I can't specify items. I can, however, advise you to bring whatever you will need to keep your offspring happy, comfortable, and occupied. You may discover, much to your chagrin, that your young children do not share your appreciation or awe for the eclipse.

☐ **POWER INVERTER** You can't plug most laptops or video players directly into a car. A small DC-to-AC power inverter will let your passengers play games or movies for the whole length of the trip without having to worry about draining the batteries in their devices. Another similar device is a car-lighter-to-USB-socket. Such adapters can run or charge items that don't require much power.

☐ **A HAT** This will keep the Sun off your head and also your neck if you choose a hat whose brim is wide enough. You may sweat, but that's a great trade-off. Keep drinking water.

☐ **A PILLOW** Actually, bring a pillow for every chair you bring. Your passengers also can use these in the car.

☐ **SUNGLASSES** Remember, despite their name, sunglasses are not for viewing the Sun. They are for providing eye comfort when you look at everything else.

☐ **CASH** Some places may not take credit/debit cards and with the huge numbers of people in transit, it may save you some time just paying with cash.

☐ **INSECT REPELLENT** The farther down the shadow's path toward the Southwest you set up, the more important this item will become.

☐ **PHONE** I list this item mainly for completeness. Does anyone ever forget his or her phone anymore? Note that, at large events (especially in smaller towns), cell towers may be overwhelmed by the number of people trying to access their cellphones. Be sure to tell anyone tracking your movements that you may be out of touch.

☐ **TELESCOPE** Be sure also to bring the minimum number of items to go with it. I won't detail them here because everyone's scope "kit" is different. Read all about telescopes in Chapter 8.

☐ **ASTROPHOTOGRAPHY GEAR** If you're going to photograph the eclipse, make sure you have the essentials you need. Keep these items together.

☐ **ODD PARTS AND TOOLS** If you have a telescope, you understand what this item means. My kit contains extra knurled knobs, an Allen wrench set, at least two each of three types of small clamps, a micro screwdriver set (I can work on eyeglasses with this), lens cleaning paper, small zip ties, hardware for any tripods I bring, extra solar filters, and, you guessed it, duct tape.

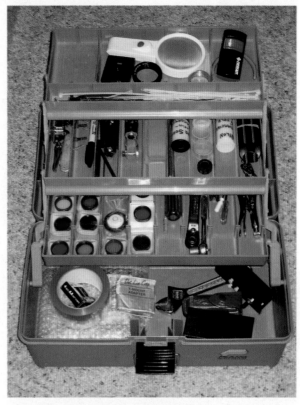

Michael's observing kit contains filters, spare parts, small tools, and more. MICHAEL E. BAKICH

☐ **PERSONAL ITEMS** You won't be hiking the Himalayas or venturing into the deep ocean aboard a submarine. You'll have room for a few extras. If there's something that's especially meaningful to you and you want to bring it along, no harm done.

☐ **A FULL TANK OF GAS** Although gas stations are plentiful in the U.S., you don't want to take the chance that you'll be in an area without one and running low on fuel. Also, if you're heading to a different location than the one you planned, probably because of the weather, you don't want to waste the time fueling up.

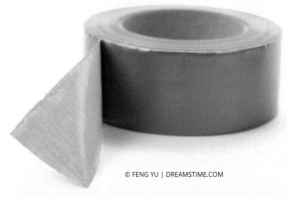

© FENG YU | DREAMSTIME.COM

Project 3
BUILD A SOLAR PROJECTOR
FROM BINOCULARS p. 52

Project 5
CAMERA CADDY p. 56

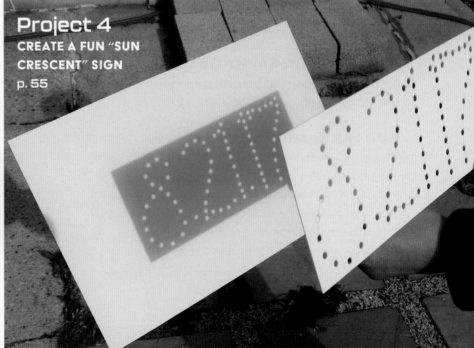

Project 4
CREATE A FUN "SUN
CRESCENT" SIGN
p. 55

DIY ECLIPSE PROJECTS

Do it yourself (DIY) projects are all over television and the internet. Through the years, and through experiencing 14 total solar eclipses (as of this writing), I've developed several that aren't difficult to complete. Look them over. You might want to try a few between now and eclipse day, and the first three don't even require an eclipse.

Project 1
HOW TO BUILD A SIMPLE SUN VIEWER

Let's say you want to view the Sun, but you don't have an approved solar filter for your telescope. Maybe you don't even have a telescope. No problem. Somewhere at home, you must have a cardboard box. That and a few common supplies will let you build a pinhole viewer that you can use to observe the Sun. While I admit that the image you'll see won't rival that through a properly filtered telescope, it will be the Sun, and you can watch it as long as you want without any safety concerns. Plus, building this viewer is essentially free.

Just follow the steps in the captions starting on the next page. Feel free to experiment with the size of the box, whether or not you leave the extra cardboard pieces attached, or the pin's diameter. In fact, if you're a clever soul, you may want to mount the box on a thin plywood base, which you can then attach to a sturdy camera tripod. That way, you won't have to hold the box while viewing the Sun. This activity is great for school classes. In that case, I suggest the teacher cuts out the hole as an additional safety factor.

1 • Start with a cardboard box at least 18 inches long. If you use a box shorter than this, the projected image of the Sun will be unacceptably small.

All photos in this chapter are by the author

2 • On one of the box's smaller ends, trace a circle by using a quarter.

3 • Carefully cut out the circle with a sharp knife.

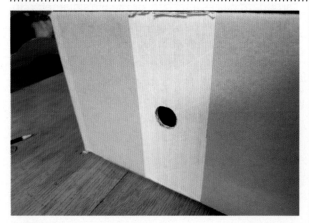

4 • Note that the hole doesn't have to be perfect — or even round!

5 • Cut out a small piece of aluminum foil big enough to cover the hole in the box. The best foil to use is a fairly thick variety.

6 • Tape the foil over the hole.

7 • Poke a hole in the foil using a pin. A pin is better than most items you could use because you want a hole with clean (not ragged) edges. As a variation, you can try using a sharp pencil. That will produce a larger hole. Larger holes let more light through and so produce a brighter Sun image. Careful, though. If your hole is too big, you'll lose the pinhole camera effect and you'll just see an out-of-focus blob.

8 • Inspect the hole for roundness and make sure nothing is blocking it.

9 • Cut away as many of the flaps as you want. Alternatively, you could fold them over and tape them to the box's sides, making the assembly a bit sturdier.

10 • Tape a white piece of paper on the inside of the box opposite the hole.

11 • Point the hole at the Sun and observe the projection on the paper. Congratulations!

Project 2

BUILD A SOLAR PROJECTOR

If building a simple Sun viewer out of a cardboard box is a bit too simple for you, or if the image quality isn't what you expected, then try this project. I guarantee you'll like the results. You'll need a little lumber, a bit of hardware, and a telescope finder scope that lets you view straight through it. (Some have a right-angle bend.)

I suggest you go one step further. Buy a bit of sandpaper (a variety pack made of several different grits) so you can give whatever wood you choose a smooth finish because no one wants a splinter.

It doesn't really matter what type of finder scope you use. They range from extremely simple to "simple but more expensive." The larger the finder scope's front lens, the more it generally costs. Now it may sound funny, but it's lucky most manufacturers of finder scopes use fairly inexpensive eyepieces. This is lucky because a complex, expensive multi-element eyepiece will trap more heat between its elements. Simple lenses will warm some but won't retain most of the longer-wavelength infrared radiation (heat).

The first measurement you'll need to make is the distance from the finder scope's eyepiece to where the Sun comes to a focus. It's really nice to have a second person's help with this. Go out and prop up a white piece of cardboard so it sits facing the Sun. You can also tape any white piece of paper to a board. Then hold the finder scope so it projects the Sun's image onto the cardboard or paper. It'll take a bit of finagling to hold steady, but it's not difficult. When you have a reasonably steady image, measure—or, better yet, have person number two measure—the distance between the eyepiece and the paper. Let's call that distance T, and here we'll measure it in inches. Feel free to use centimeters, cubits, or whatever you like. While you're at it, measure the diameter of the Sun's projected image. We'll call that measurement D.

The long narrow board you will mount everything to, which I call the long arm, must be T inches long plus enough extra length to accommodate the finder bracket and the thickness of the square piece of wood you'll use to project the image onto. I call that piece the screen board.

For the finder scope that I chose, T equals approximately 29¾ inches (75.6 centimeters) and D equals 4¼ inches (10.8 cm). The extra length needed to attach the finder scope is 7 inches (17.8 cm). I chose a screen board ¾-inch (1.9 cm) thick, and I suggest you do the same because you'll need to attach it to the long arm with two screws. Thinner boards may crack. Also, because D equals 4¼ inches, I made the screen board a 6 x 6-inch (15.2 x 15.2 cm) square. With a tiny bit of extra length on each end, this made my long arm 38 inches (96.5 cm) long. I chose to make that piece 2¼ inches (5.7 cm) wide, and the board already was ¾ inch thick. I chose that width because it was just a bit more than I needed to accommodate the base of the finder scope. Any additional width would just add useless weight.

Next, sand the long arm to a smooth finish so there won't be any splinters. Then attach the finder scope base using two 1-inch-long bolts. Each bolt is long enough to go through the finder's base, but not long enough to stick out from the bottom of the long arm. Once complete, drill two holes wide enough to accommodate lock washers and nuts attached to the bolts and deep enough to keep

2 • I mounted the finder scope bracket on one side of the long arm, which I made out of cherry and sanded to a smooth finish. No splinters here.

1 • To make the solar projector, I used an 8×50 finder scope that was lying around.

them from protruding. Then attach the screen board to the long arm with two 1½-inch-long (3.8 cm) drywall screws. Make sure it's perpendicular to the long arm and parallel with the finder scope's front lens.

The next step is drilling the hole that the tripod's bolt will screw into, securing it to your tripod. Don't just measure to the center of the long arm and drill. Rather, attach the finder scope base, the finder scope, and the screen board to the long arm. Then, carefully balancing the long arm on your two index fingers held about 30 inches (76 cm) apart, slowly move your fingers toward each other. They will meet at the point where the long arm's weight balances. Mark the spot, drill a hole, and tap it for a ¼-20 thread. Specifications say you should use a 13/64-inch bit, but when I'm tapping wood I use a slightly smaller one, either 11/64-inch or 3/16-inch. After you attach a piece of white paper to the screen board, you should be ready to rock. Choose a tripod sturdy enough to carry the weight and hold the arm steady, and attach your solar projector to the tripod.

A couple of final notes. First, if you have lens caps for your finder, use them. They will keep dirt off the optics. No caps? Use two small kitchen sandwich bags secured with two rubber bands. Second, when you're transporting the solar projector, disconnect the finder scope and put it somewhere it won't get abused. Third, don't over-tighten the tripod-mounting bolt. Remember, you're screwing it into wood. If you do strip the threads, home supply stores sell metal ¼-20 inserts you can attach (usually, you hammer them into place), which you can screw the tripod's bolt into.

3 • The screen board attaches to the end of the long arm opposite the finder scope. The board is ¾ inch thick and attaches to the board with two screws.

4 • The holes beneath the finder scope's mounting bracket are large enough to accommodate the screw and nuts without them protruding beyond the bottom of the long arm.

5 • The finder scope used for this solar projector focuses when you loosen the knurled knob, move by rotating the front lens some, and then re-tighten the knob.

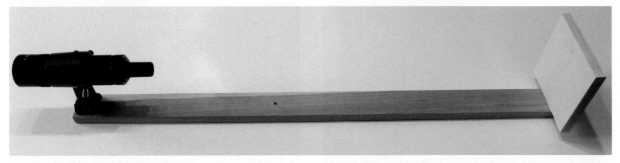

6 • The completed solar projector is simple, disassembles easily, and works like a charm. Note that the tripod's mounting hole is nowhere near the center of the board's long dimension. That's because the finder scope and its base weigh significantly more than the screen board.

Project 3

BUILD A SOLAR PROJECTOR FROM BINOCULARS

If you're not an amateur astronomer, you may not have a finder scope lying around to construct Project 2. It's likely, however, that you have binoculars, possibly some that you haven't used in years. Here's how to make the same kind of viewer using them.

For this project, I personally chose not to use any of the binoculars I own. All are high-quality and expensive, and I didn't relish the thought of sending unfiltered sunlight through the multiple optical elements within each unit. Instead, I ventured online and spent $24 for the 7×35 binoculars pictured in this project. They have a much less complex optical path than the ones I own. And you know what? They work great!

Something to remember, if you try this project, is that most of you will be mounting the finished product on a camera tripod. So, try to save weight wherever possible. I strove to make this unit as small as practical and still provide the support the binoculars need. I also performed a totally unnecessary step. I drilled a number of 1-inch diameter holes along the base's length. This step will reduce the unit's weight without sacrificing much strength.

The parts that make this projector are a wooden base, a metal L bracket, a metal plate the tripod bolt can screw into, and a wooden support for a screen. The binoculars I used measure 7¼ x 5¼ inches, but you don't need to make the base 5¼ inches wide. It's only there to provide attachment points for the L bracket and the camera tripod.

My base is 2 inches wide and 26 inches long. I estimated its length by letting the Sun's light pass through the binoculars and measuring how far from the eyepiece the image focused. For this unit, that focused length was 20 inches. That's the distance I determined the Sun's image was large enough to easily see, but not so large that it became faint. With this in mind, the base had to be

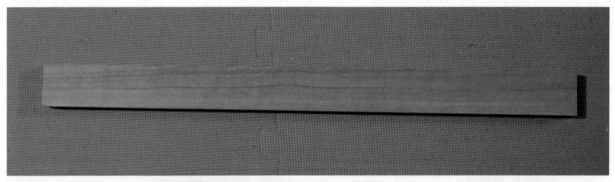

1 • The base of my binocular projector is a piece of cherry that measures 26 by 2 by ¾ inches.

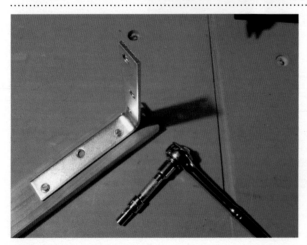

2 • I used two wood screws to attach the L bracket to the base. My L bracket measures 6 x 6 inches.

3 • Nearly all binoculars have a ¼-20 threaded hole to allow you to attach a bracket.

4 • The backstop is a piece of ¼-inch-thick plywood 8 inches on a side. I attached it to one of the base's ends with two wood screws.

5 • With everything attached, I found the unit's balance point by slowly rolling it back and forth on a pencil.

6 • The threaded hole of the tripod-mounting plate lies at the balance point of the binocular projector's long axis.

a bit more than 5¼ plus 20 inches long.

I mounted the L bracket to the base so the ends of the binocular's front lenses were even with the front end of the base. I didn't want any of the base sticking out past the binoculars. I used wood screws to secure the L bracket to the base and a short ¼-20 bolt to attach the binoculars to the bracket. If you use a lock washer with the bolt, the bolt will be less likely to loosen, so your binoculars will be less likely to shift.

The only consideration for the L bracket is that it holds the binoculars high enough so the Sun's image isn't cut off at the base. The easiest way to figure out the height you'll need is to first measure the diameter of the Sun's focused image (that you created two paragraphs ago). Then take one-half that length and add ½ inch to it.

I then attached the 8 x 8-inch, ¼-inch-thick piece of plywood that would serve as the backstop to a white piece of thin cardboard onto which the binoculars will project the Sun's image. A couple of small wood screws will secure it.

The next step is attaching the plate that the tripod's bolt will screw into. I used a 1-inch-wide x ¼-inch-thick piece of steel. But you won't be attaching this at the middle of the base's length. Rather, attach the binoculars and the screen board to the base first. Then, carefully balancing the base on your two index fingers with one at each end of the base, slowly move your fingers toward each other. They will meet at the point where the base's weight (with everything attached) balances. Mark the spot, also centered according to the base's width. That's where to place the centered tripod-mounting hole in the 1 x ¼-inch bar. In fact, you should see your mark at the bottom of the hole. Remove the binoculars, L bracket, and screen board, and attach the metal bar.

If you're a bit unsteady or unsure about manipulating the base on your fingers with everything attached, there's another way you can find its center point. Make a fulcrum from a wooden dowel, a short piece of electrical conduit, or some other long round object. Its diameter can be anything from ¼-inch to 2-inches. Then set the base, with everything attached, onto the fulcrum. Move the base until you find where it balances, and make a mark.

7 • **I use a white piece of paper or thin cardboard as a projection screen. Two spring clips hold it in place.**

Once you assemble the unit and mount it on a tripod, it's your choice whether to use one side of the binoculars or both if the images don't overlap. If you're using just one side of the binoculars, the easiest way to eliminate the other side is to leave the rubber/plastic caps on it. I like the effect of having two Suns visible.

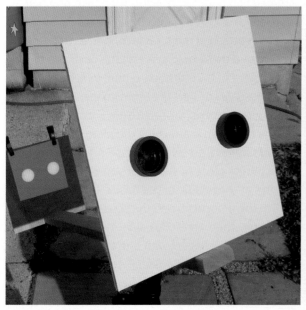

8 • **To increase the contrast of the projected Suns, I cut a piece of foam core board that fits over the front of the binoculars. Using this shield means the whole projection screen will be in shadow.**

VIDEO PROJECTS

Many of today's high-end digital cameras, and also not-so-high-end point-and-shoot cameras, include the option to record video. That opens them up to a wide array of uses during the April 8, 2024, total solar eclipse. The best thing (well, after the results you'll get) is that all you have to do is acquire some easy-to-find items and make some simple preparations. Then just hit the Record button and walk away.

For projects 6, 7, and 9, in addition to the camera, you will also need a tripod. But if you plan to do any multiples of these projects, you don't necessarily need one tripod per camera (although you can choose to set it up that way). If you follow my suggestion and build Project 5 first, you'll need only one tripod.

Project 4
CREATE A FUN "SUN CRESCENT" SIGN

For this project, all you need is some white poster board and something to poke small holes in it. Lacking a poster board, you also can use comic book backing boards. Each of these choices is thin enough to allow you to punch holes, but sturdy enough not to bend in mild breezes.

Do note that the cleaner your holes are the better your final image(s) will look. To that end, I purchased a cheap set of six leather punches online. They're all quite sharp and produce exceptionally clean holes. But which size to use?

This was easy enough to test. I made three holes with each of the six punches in the set. I then took the card I had punched outside with another white card and studied the Sun's projected images. I picked the one that looked the best to me.

After you perform these easy tasks, it's time to let your creative juices flow. Cut a piece of poster board roughly 7 x 3 inches. Then draw a pattern. Most people use a word, a name, a place, or the date of the eclipse. Mark dots often enough along your lines that you can recognize the word/date just by looking at the dots. Then, by using your punch, turn each dot into a hole.

On eclipse day, as the Moon crosses the Sun's disk, the images will turn into crescents. Your word will be projected onto a slightly larger piece of poster board. Have someone photograph you holding it. Below is a design I made for the 2017 eclipse.

And here's another thought. As desperately as I want all of you to experience totality, I understand that some people will not be able to stand under the Moon's umbra. The previous four projects will still work, however. Each will show a partially eclipsed Sun, which, really, is also what they will show for people in the path of totality, too. That said, Project 4 will show the best results if the Moon's coverage of the Sun is 75 percent or greater.

1 • To test the size of the hole I wanted, I punched a series of various sizes into a piece of cardboard.

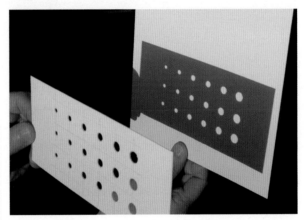
2 • I took my test card outside and let the projected images of the Sun help me pick the right size hole.

3 • For your sign, create a pattern meaningful to you. Make it as simple or as complex as you like.

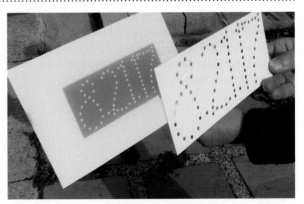
4 • The finished card lets you project multiple images of the Sun onto a second piece of white cardboard.

Project 5
CAMERA CADDY

With the availability of inexpensive new as well as used point-and-shoot cameras, it's possible you may wish to carry out more than one of the following video projects. You can even invent your own. My thought was that I had invested enough in the cameras, so I didn't want to buy multiple tripods. Other considerations come into play besides the cost, of course: storage, transporting them to the eclipse site, carving out enough territory to set everything up, and ease of access.

To simplify my video projects, I came up with a simple device I made from materials lying around my shop coupled with an inexpensive online purchase. Here's how I did it.

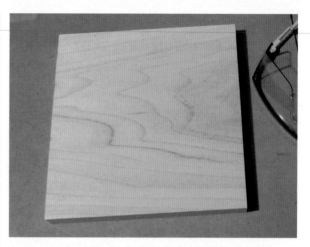

1 • **I started with a board 7½ inches square and ¾ inch thick. I used a piece of hardwood because I like the feel of it after sanding. Plywood, however, would work equally well.**

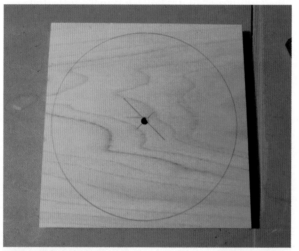

2 • **I marked the center of the board and, using a compass, scribed a circle. I didn't actually measure the circle at the time, but it turned out to be approximately 6¾ inches in diameter. NOTE: You can leave the board square. I cut the edges off because I wanted to get rid of as much extra weight as possible and because I think it looks a bit more finished. Go with your preference. After drawing the circle, I drilled a ¼-inch hole in the board's center.**

3 • **I then used a band saw to cut along the circle.**

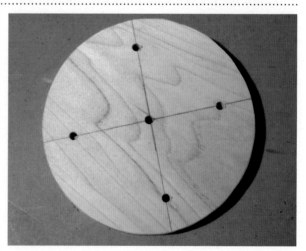

4 • **After extending the initial centering marks, I drilled four ¼-inch holes, each 1 inch in from the edge.**

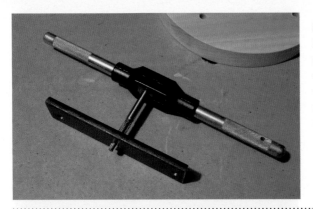

5 • I then found a piece of 1-inch-wide x ¼-inch-thick bar stock (you can purchase this at any home supply store). I cut mine about 4¾ inches long. I drilled and countersunk two mounting holes near each end and then drilled a hole that I tapped to ¼-20 size. Charts recommend a no. 7 drill for this purpose, or you can use the nearest fractional equivalent, 13/64 inches.

6 • To hold the cameras, I purchased four (actually, I bought six) mini-ball-head holders. Each of these measure only 2½ inches long and have a ¼-20 male end that screws into the bottoms of the cameras. I found them on eBay for $1.98 each, and shipping was free.

7 • This picture shows the bottom of the Camera Caddy. I sanded each edge of the metal bar and the wooden base so I won't get an injury if I have to move fast on eclipse day.

8 • And here's the top of the Camera Caddy with all four mini-ball-head mounts attached.

This last picture shows the completed project mounted atop a camera tripod.

Project 6
RECORD THE DARKENING

This is the simplest video project you can perform, and as a bonus, it requires no additional equipment besides a tripod. I have set this up for the past half dozen or so total solar eclipses I've experienced, and afterward, I always was glad I did.

First, determine your "spot" for viewing the eclipse. It may be with your spouse, family, friends, or telescope, but you'll need to stake out an area before the eclipse begins.

Move your camera on its tripod far enough away from where you're set up so that it frames you completely. From this point onward, the camera will be facing directly away from the Sun's position at mid-eclipse.

At a specific time you've chosen in advance, begin recording the video. I have used either 15 or 10 minutes prior to totality. Then just let the camera run until an equal amount of time has elapsed after totality, and stop recording.

You will then have a video record of the darkening (and subsequent re-brightening) that occurred during the most dramatic part of the eclipse. In addition, if your camera also records sound when in video mode (and pretty much every camera does), you will have captured your reactions, and probably those of people around you, to watch or share with others whenever the mood strikes you.

I have three important notes to add to these instructions. First, be sure your camera isn't set to compensate for decreasing light levels. You want the scene to darken. If it does adjust for darkening, the model's instruction manual will reveal how to cancel it.

Project 7
RECORD THE SHADOW PASSING OVERHEAD

This project is just as simple as no. 6. First, however, there must be clouds at your location. That's not something I will wish on you. In fact, I hope your location has blue sky from horizon to horizon. If broken clouds (usually cumulus) exist, however, or if you can see distant clouds on the horizon, you may want to record the passage of the Moon's shadow as it speeds toward your location.

Simply point the camera in the direction of the approaching shadow, if clouds lie in that direction, or aim it in the direction the shadow travels after it covers your location if you see clouds there. Then follow the instructions for Project 6. For the April 8, 2024, eclipse, the Moon's shadow will race across the U.S. from the southwest (Texas) to the northeast (Maine).

Second, make certain the chip in your camera has enough memory to record 20 to 30 minutes of video. Some cameras will display the number of "minutes left," on the display, or some such message, when you begin video recording. Others provide this information in the instruction manual, which will say something like, "If you use a 16Gb memory card, you can record XX minutes of video." If your camera does neither, then charge the battery, hit video record, and see how long it will go before your memory card fills up. Actually, if it's still going after half an hour, you can stop. That's long enough.

Finally, make absolutely sure the battery you're using to record video is fully charged before you begin, and the spare battery too. This detail is easy to overlook.

Project 8
RECORD THE TEMPERATURE DROP

For this exercise, you'll need a digital thermometer and a digital watch or clock. Actually, an analog thermometer that contains mercury will work, but the temperature readings won't be as precise. Likewise, an analog watch or clock will work, but discerning the time won't be as quick. Please note: If you go the digital thermometer route, choose one that doesn't have a timer to shut its display off after a certain time; 10 minutes is the usual amount.

One plus related to this project is that you don't need a tripod. You can set the three components on a table, a board, or, in a pinch, directly on the ground. If you have a way to shade the thermometer from direct sunlight, the temperature reading will be more accurate.

This last tip will apply to only some of you. If your camera has an intervalometer, an internal timer that trips the shutter at pre-set intervals, use it for this project. You really don't need a continuous

You can record the temperature drop at your location by taking a video of a setup like this one.

reading of temperature because it won't change that fast. If, however, you can set your camera to record a frame every 1, 2, or 5 seconds, the movie you create will be 1/30, 1/60, or 1/150 the size of a continuous one at 30 frames per second, and it will playback that much faster.

THE PATH OF THE 2024 ECLIPSE

The April 8, 2024, total solar eclipse touches only three countries: Mexico, the United States, and Canada.

PACIFIC OCEAN

From the start of the eclipse until it touches the mainland of Mexico, the path crosses more than 4,000 miles of the Pacific Ocean. The only land it touches along the way is four tiny islands. The first is Isla Socorro (Socorro Island), which lies 370 miles off Mexico's western coast. From Isla Socorro's southeast corner, which is closest to the center line, totality will last 3 minutes 33 seconds with the Sun 65° high in the east-southeast.

The next contact with land occurs much closer to Mexico at three of the four islands of the Islas Marías (Mary Islands). Isla San Juanito and Isla Maria Madre, the only populated island with some 1,100 residents, lie completely within the path of totality, while Isla Maria Magdalena is bisected by its southern limit.

MEXICO

The Moon's umbra touches the coast of Mexico at 12:07 p.m. Mexican Pacific Daylight Time less than 12 miles southeast of Mazatlán. The southeastern suburbs of this city will enjoy 4 minutes and 20 seconds of totality. As far as sizable cities, Durango is next. It will enjoy 3 minutes 50 seconds of totality with the Sun 70° high.

At 1:15 p.m. Mexican Central Daylight Time, the umbra arrives at Nazas. This town of some 4,000 residents will surely see that number swell because it's the nearest location to the point of longest totality. The town itself will enjoy that duration—4 minutes 28 seconds—but the actual spot is about 3 miles to the north.

Another ideal destination is Torreón. This city, with a population in excess of 1.5 million, can handle a large influx of eclipse chasers. The city center will receive 4 minutes 14 seconds of totality

Monterrey will host many travelers who want to view the eclipse. DANIEL ESCOBEDO/WIKIMEDIA COMMONS

with the Sun 70° high at mid-eclipse. But head just 25 miles to the northwest, perhaps along Highway 490 about 3 miles south of Bermejillo, and you'll gain back the 14 seconds that will allow you to enjoy the greatest duration of totality. If the road is crowded or you're a bit unsure whether you should park, head into town and you'll lose only 1 second off the maximum.

A bit more to the northeast, Monterrey will be a launching point for eclipse chasers. With a population of 4.8 million, it holds every amenity a tourist could ask for, especially easy access to totality. The city itself will experience a 95-percent-partial eclipse, but the real drama takes place to the northwest.

The Alamo Mission in San Antonio is one of the most visited historic sites in the U.S. RENELIBRARY/WIKIMEDIA COMMONS

The southern limit of totality lies 110 miles away, and the center line is 62 miles farther than that. Take Highway 53 to Monclova (2 minutes 27 seconds of totality), and then Highway 30 to Cuatro Ciénegas, where you'll enjoy 4 minutes 17 seconds of totality. For an additional 10 seconds, drive 20 miles northwest on Highway 20, the road to Ocampo.

Another large city, one that lies across the Rio Grande River from Eagle Pass, Texas, is Piedras Negras. Anyone from the U.S. who wants to experience the maximum duration of totality (4 minutes 28 seconds), will either cross the Eagle Pass–Piedras Negras International Bridge or the Camino Real International Bridge. Using Piedras Negras as a base is a good idea because its metro population is a quarter million.

TEXAS

Totality will arrive at Las Quintas Fronterizas, Texas, at 1:27 p.m. CDT. This town of some 2,500 residents will experience 4 minutes 22 seconds of totality, but a brief 12-mile drive north along U.S. Route 277 to tiny Radar Base will gain eclipse watchers an additional 5 seconds of umbral darkness. Only a few hundred residents live in this community.

As the umbra begins its northeast journey, it touches quite a few Texas towns. Leakey, with a population of 650, and the unincorporated community Concan lie near the center line as well as along

If April 8, 2024, is clear, millions of people in Dallas can just look up and view the eclipse. MATTHEW T. RADER/WIKIMEDIA COMMONS

U.S. Hwy. 83. Leakey will receive 4 minutes and 25 seconds of totality, and Concan gets 1 second less.

The first significant city in the U.S. the umbra will touch is San Antonio. With a metro area containing 2.5 million residents, easy access, numerous attractions, and some of the best weather prospects in the country, San Antonio will be a starting point for eclipse chasers from around the globe. The path's southern limit splits the city.

Travelers may choose to take Interstate 10 northwest out of San Antonio to Comfort (4 minutes 13 seconds of totality), Kerrville (4 minutes 24 seconds), or Ingram (4 minutes 26 seconds), which lies on the center line. Fredericksburg, another possibility 19 miles farther north, enjoys 4 minutes 24 seconds beneath the umbra.

The next large city that will experience totality is Austin. From the state capitol, lawmakers will enjoy 1 minute 59 seconds of totality with the Sun 67° high. Photographers could capture the eclipsed Sun above the Capitol, but an extra 2 minutes 25 seconds of totality lies just 56 miles to the northwest via State Highway 71.

After passing over the mid-size cities of Killeen (4 minutes 18 seconds of totality) and Waco (4 minutes 14 seconds), at 1:40 p.m. CDT, the Moon's umbra will encounter the outskirts of Dallas-Fort Worth. Students can expect 2 minutes 35 seconds of totality at the Amon G. Carter Stadium at Texas Christian University. At AT&T Stadium in Arlington, where the Dallas Cowboys play their home football games, totality will last 3 minutes 17 seconds.

As the umbra is encountering Waco and Killeen, several smaller towns with slightly longer durations of totality lie along the center line. Lampasas, Topsey, Gatesville, and Valley Mills all will enjoy 4 minutes 24 seconds of umbral darkness.

After encountering Dallas-Fort Worth, the umbra continues its advance to the northeast. The best locations prior to the Oklahoma border are the towns of Sulphur Springs and Clarksville. Residents and visitors to these locations will enjoy 4 minutes 20 seconds of totality. Indeed, the Lone Star State, with its large cities and their amenities, excellent roads within the path, and the best weather prospects in the U.S., will be the major hub of eclipse activity in 2024. More than 12.5 million Texans already live within the path totality.

OKLAHOMA AND ARKANSAS

Totality begins in Oklahoma and Arkansas simultaneously at 1:45:39 p.m. CDT, at the point where the two states meet Texas at the Red River. Anyone at that location will experience 4 minutes and 5 seconds of totality. The umbra will cover only a small part of the southeast corner of Oklahoma, but the towns of Idabel (7,000 residents, 4 minutes 19 sec-

Science lovers who base in Little Rock, Arkansas, might want to visit the Museum of Discovery.

LITTLET889/WIKIMEDIA COMMONS

onds of totality) and Broken Bow (4,000 residents, 4 minutes 16 seconds) are well placed for the event. And although the towns of Haworth and Eagletown also get 4 minutes 19 seconds of darkness, they contain only about 300 and 500 residents, respectively, so the amenities in them will be limited.

Cape Girardeau lies 100 miles southeast of St. Louis, on the bank of the Mississippi River.

U.S. ARMY CORPS OF ENGINEERS

As the dark shadow moves through this region, it just misses some large metropolitan areas: Shreveport, Louisiana (98-percent partial eclipse), Oklahoma City (94 percent), and Memphis, Tennessee (98 percent). It will, however, bring 3 minutes of totality to the western parts of Arkansas' capital city of Little Rock, which will certainly be a base for many people.

Arkansas offers numerous small towns along or near the center line that boost the duration of totality more than a minute above that in Little Rock. A few are Gillham (4 minutes 19 seconds of totality), Umpire (4 minutes 17 seconds), Big Fork (4 minutes 18 seconds), Pencil Bluff (4 minutes 18 seconds), Onyx (4 minutes 17 seconds), Plainview (4 minutes 17 seconds), Wilson and Lost Corner (4 minutes 16 seconds each), Shirley (4 minutes 15 seconds), and Dalton (4 minutes 13 seconds).

For eclipse chasers who want the easiest possible travel route, the Interstate 40 corridor offers Russellville (29,000 residents, 4 minutes 11 seconds of totality), Pottsville (3,000 residents, 4 minutes 14 seconds), Atkins (3,000 residents, 4 minutes 16 seconds), and Morrilton (6,600 residents, 4 minutes 12 seconds). Travelers thinking about heading to any of these locations should check well in advance to see if there are any eclipse-based activities prior to the event.

MISSOURI AND TENNESSEE

The center line of totality comes to a number of small towns in southeastern Missouri. The largest, easily accessible from Interstate 55, is Cape Girardeau, a city of some 80,000 residents. If the sky is clear on eclipse day, they will experience 4 minutes 6 seconds of totality with the Sun 57° high at mid-eclipse. Or, they can drive about 10 miles west on State Route 72 to Jackson for an additional 4 seconds of darkness. Two smaller communities, Doniphan and Zalma, will experience 4 minutes 12 seconds and 4 minutes and 11 seconds of totality, respectively.

Technically, Tennessee can claim to be one of the states touched by totality. However, only a tiny part of its northwest corner is covered by the umbra. If you must see totality from this state, head north from Tiptonville along State Route 22 to the banks of the Mississippi River, which, because of the severe bend in the river, will be both to your east and west. At the Kentucky border there, totality will last 1 minute 25 seconds with the Sun 58° high at mid-eclipse.

KENTUCKY AND ILLINOIS

Only a small part of far-western Kentucky experiences totality. The largest town in the state covered by the umbra is Paducah, whose 25,000 residents

Saluki Stadium on the campus of Southern Illinois University will host thousands of eclipse watchers if the sky is clear. DEVOUT1145/WIKIMEDIA COMMONS

If you view the eclipse from Indiana, be sure to take the family to the Children's Museum of Indianapolis.

will see 1 minute and 40 seconds of totality. Travelers headed to (or from) there will find easy access because of Interstate 24.

It will be far better for them to make the 50-mile drive to Carbondale, Illinois, or a town near it. That city, home of Southern Illinois University, was a hub of activity during the August 21, 2017, total solar eclipse, and surely will serve that purpose again. In 2024, however, the duration of totality increases to 4 minutes 9 seconds, up almost a minute and a half from 2017.

Mt. Vernon certainly will be a busy pass-through point around April 8, lying as it does at the intersection of Interstates 57 and 64. This city of some 20,000 will experience 3 minutes 38 seconds of totality.

Other Illinois localities on or near the center line include Johnston City (4 minutes 8 seconds of totality), Thompsonville (4 minutes 8 seconds), McLeansboro (4 minutes 8 seconds), Springerton (4 minutes 7 seconds), Albion (4 minutes 7 seconds), Friendsville (4 minutes 6 seconds), and Lawrenceville (4 minutes 2 seconds). In each of these places, the Sun will stand some 55° above the horizon.

Illinois' great city, Chicago, lies well north of the eclipse path. The many residents there will experience a 94-percent-partial eclipse. Those interested in totality can drive 320 miles south along Interstate 57 to Marion for 4 minutes 9 seconds of totality, or 235 miles to Bloomington, Indiana, where totality will last 4 minutes 3 seconds.

INDIANA

As totality progresses to the northeast, it barely misses three more large cities: Nashville, Tennessee (95 percent), St. Louis (99 percent), and Louisville, Kentucky (99 percent).

The umbra will cover a wide swath of Indiana, but most of the attention will focus on the state's capital city. Downtown Indianapolis—a metropolis served by four interstate highways—will get 3 minutes 46 seconds of totality. That said, a quick 22-mile drive southeast to Fairland will boost the duration to 4 minutes 2 seconds. Other smaller, but still favorable, locations in this state include Vincennes (4 minutes 6 seconds of totality), Bloomington (4 minutes 2 seconds), and Muncie (3 minutes 43 seconds).

As the umbra continues past Indianapolis, the center line intersects several more small communities. Among them are Morristown, Carthage, Knightstown, and Spiceland, all with 4 minutes 1 second of totality, and Economy and South Salem, both of which receive an even 4 minutes.

The Cincinnati Observatory is a wonderful astronomical facility that will host activities days before the eclipse.

Music aficionados visiting Cleveland will want to check out the Rock and Roll Hall of Fame. MUSIKANIMAL/WIKIMEDIA COMMONS

OHIO

When the umbra moves through Ohio, it covers—completely or partially—that state's four largest cities: Columbus, Cleveland, Cincinnati, and Toledo. Of these, Cleveland will get the best of it with a duration of totality downtown of 3 minutes 50 seconds with the Sun 49° high.

Unfortunately, most of Cincinnati lies outside the path of totality. Residents there will probably drive northwest about 80 miles to Shelbyville or Knightstown in Indiana, both of which will enjoy 4 minutes of umbral darkness.

Ohio's capital, Columbus, will probably see a mass exodus of eclipse chasers in the days prior to the event. Only in the city's northwest suburbs is the eclipse total, and not for very long.

Unlike Columbus' near-miss with the Moon's umbra, most of Toledo lies in the path, although it's at the northern limit of totality. Downtown there will get 1 minute 50 seconds of darkness. Eclipse chasers there would be well

Erie's Presque Isle State Park will host many eclipse watchers if the sky is clear April 8, 2024. RUHRFISCH/WIKIMEDIA COMMONS

advised to head east about 70 miles on Interstate 80/90 to the center line at either Florence or Birmingham, each of which will enjoy 3 minutes 54 seconds of totality.

Finally, around 2 p.m. EDT, the umbra arrives in Cleveland. Totality in the metro area will last 3 minutes and 50 seconds. An additional 3 seconds can be gained by driving about 30 miles west along U.S. Highway 20—which turns into U.S. Hwy 6—to the center line at Avon Lake. To enjoy a totality of 4 minutes, however, a traveler would have to drive some 250 miles to South Salem, which lies on the center line northwest of Dayton.

MICHIGAN

Like Tennessee, Michigan will receive a minuscule amount of totality. Across roughly 50 square miles in the state's southeastern corner, where U.S. Hwy. 24 and Interstate 75 cross its border with Ohio, viewers will experience a maximum duration of 1 minute with the Sun 50° high in the southwest. Unfortunately,

Totality will last 3 minutes 45 seconds at the Buffalo Museum of Science. PUBDOG/WIKIMEDIA COMMONS

Detroit lies outside of the path, although it gets a 99.4-percent-partial eclipse along the banks of the Detroit River.

PENNSYLVANIA

Totality begins in Pennsylvania at 3:15:38 p.m. EDT, just south of where Interstate 90 crosses the border from Ohio. Anyone at that point will enjoy 3 minutes 44 seconds of totality with the Sun 48° high. Just east of there, the small town of West Springfield, which has a population of around 3,300, receives 1 second less beneath the umbra.

The largest city in the state to enjoy totality will be Erie. Residents in the downtown area will be treated to 3 minutes 42 seconds of totality. Numbers here surely will swell on eclipse day because the great city of Pittsburgh lies only 128 miles south and eclipse chasers who don't drive to Cleveland may choose instead to make the easy drive to Erie via I-79.

The longest duration of totality in the state will occur along the shore of Lake Erie. Fortunately, this 45-mile stretch is paralleled by I-90, so access is not difficult.

ONTARIO

Almost simultaneously with touching Michigan, at 3:14 p.m. EDT, the umbra enters Ontario. The first large city covered by the umbra is Hamilton. Any of its residents can experience 2 minutes 15 seconds of totality from the southeastern part of the city. There, the Sun will stand 46° above the southwestern horizon. St. Catharine's, only 25 miles to the east, will add a minute to that duration from its southeastern reaches.

Unfortunately, Canada's largest city, Toronto, lies just outside the path of totality. Residents who remain there on eclipse day will see the Moon cover 99.9 percent of the solar disk.

Most eclipse chasers in Toronto, however, probably will head for Hamilton. Or they'll continue farther south to the shore of Lake Erie. If it's clear on eclipse day, a great view will come from Long

Hamilton, Ontario, Canada, will be under the Moon's shadow on eclipse day. © ANJELAGR | DREAMSTIME.COM

Niagara Falls will be one of the most picturesque sites from which eclipse chasers can view the event.

Point Provincial Park. Totality there will last 3 minutes 42 seconds with the Sun 47° high in the southwest. Other residents throughout this province who want the maximum amount of time beneath the umbra should head to Fort Erie, which lies at Ontario's border with New York. At the shoreline south of this town of 30,000, totality will last 3 minutes 45 seconds.

NEW YORK, VERMONT, AND NEW HAMPSHIRE

The next big city that will experience the Moon's umbra is Buffalo, New York. The center line passes right through downtown, bringing this city 3 minutes 46 seconds of totality. Rochester, on the shore of Lake Ontario, also is well-positioned for totality. From there, the Moon will cover the Sun for 3 minutes and 40 seconds.

The umbra next covers upstate New York and the northern parts of Vermont and New Hampshire before entering Maine. No sizable cities lie in this region. Montpelier, the capital of Vermont, has only 7,500 residents who, if it's clear, will enjoy 1 minute 42 seconds of totality with the Sun 40° high in the southwest.

If the Northeastern U.S. has good weather on eclipse day, the most picturesque images of the event might come from Niagara Falls. From the outlook called Terrapin Point, an eclipse watcher will enjoy 3 minutes 28 seconds of totality. Almost as important, the Sun will hang 46° high in the southwest—directly over the Falls!

QUÉBEC

When the umbra enters the province of Québec, it touches Montréal. In fact, it splits the country's second-largest city. Any of the residents in the metro area that want as much totality as this city gets should head to the southernmost point on the banks of the St. Lawrence River. There the Moon will cover the Sun for 1 minute 57 seconds 40° above the southwestern horizon at mid-eclipse.

Other spots in Québec closer to the center line feature more totality. These include Sherbrooke (3 minutes 22 seconds, 43 miles east of Montréal) and Lac-Mégantic (3 minutes 26 seconds, 68 miles east of Montréal).

MAINE

As the umbra heads back into the U.S., it misses the largest of Maine's cities because they lie south of the path of totality. That said, all are along Interstate 95, which leads north to Oakfield or one of the small towns near it. There, totality will last 3 minutes 23 seconds with the Sun 36° high at mid-eclipse.

Totality leaves the United States for good in 2024 at 3:35 p.m. EDT on the eastern edge of Littleton, Maine. An observer there would enjoy 3 minutes 22 seconds of totality with the Sun 35° high in the west-southwest at mid-eclipse.

NEW BRUNSWICK, PRINCE EDWARD ISLAND, AND NOVA SCOTIA

After reentering the U.S. and crossing Maine, the umbral shadow touches New Brunswick at 4:32 p.m. Atlantic Daylight Time at Bloomfield. There, totality will last 3 minutes 21 seconds with the Sun still 35° high. As it leaves this province, totality comes to half of Prince Edward Island, all of Grindstone Island, and the northern tip of Nova Scotia. At

If the weather cooperates, New London, Prince Edward Island, in Canada would be picturesque. RIXIE/DREAMSTIME.COM

Meat Cove, totality will last 1 minute 42 seconds with the Sun 30° high in the west-southwest at mid-eclipse.

NEWFOUNDLAND AND LABRADOR

The last solid ground to enjoy totality in 2024 is the island of Newfoundland, which is part of Newfoundland and Labrador, Canada's easternmost province. Specifically, that point will be the small town of Catalina, located on the eastern side of the Bonavista Peninsula. There, totality will last 2 minutes 53 seconds with the Sun 23° high in the west-southwest.

ATLANTIC OCEAN

After crossing the Canadian coastline, the umbra heads out over the North Atlantic Ocean for the final 2,500 miles of its journey. It won't touch the contiguous U.S. again for another 20 years, so be sure to experience totality in 2024.

The Moon's umbra last touches land at the Bonavista Peninsula in Newfoundland. TANGO7174/WIKIMEDIA COMMONS

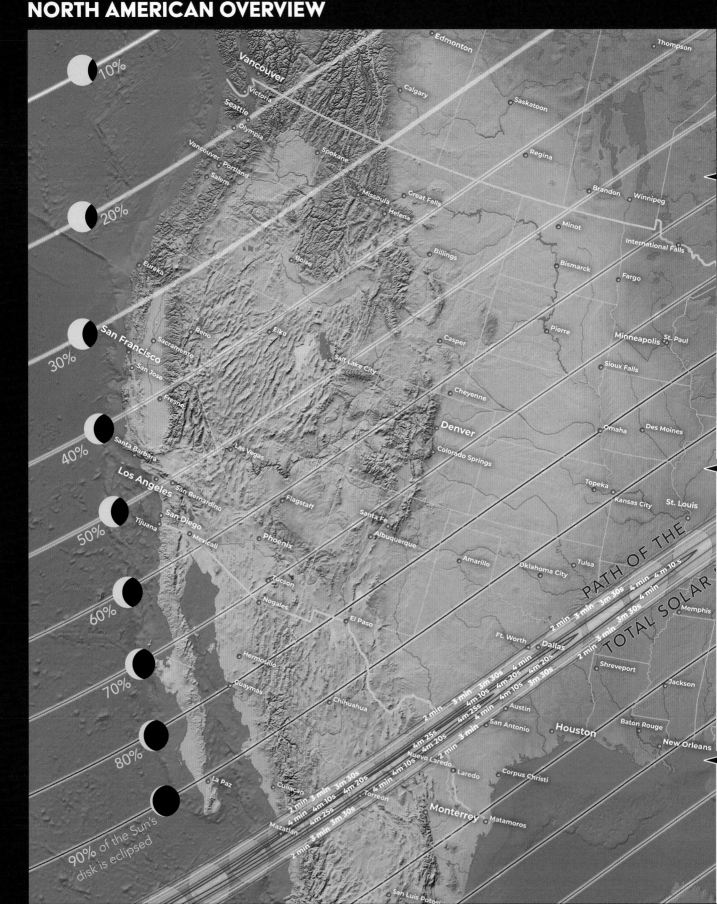

10%

20%

San Francisco

30%

40%

Santa Barbara

Los Angeles

50%

60%

70%

80%

90% of the Sun's
disk is eclipsed

Edmonton
Thompson
Vancouver
Victoria
Calgary
Saskatoon
Seattle
Olympia
Regina
Vancouver Portland
Spokane
Brandon Winnipeg
Salem
Missoula Great Falls
Minot
Helena
International Falls
Boise
Billings
Bismarck
Fargo
Eureka
Reno
Elko
Pierre
Minneapolis St. Paul
Sacramento
Salt Lake City
Casper
San Jose
Sioux Falls
Fresno
Cheyenne
Denver
Omaha Des Moines
Las Vegas
Colorado Springs
Tijuana San Diego
Topeka
Mexicali
San Bernardino
Flagstaff
Santa Fe
Kansas City St. Louis
Phoenix
Albuquerque
Tucson
Amarillo
Oklahoma City Tulsa
Nogales
El Paso
PATH OF THE
Hermosillo
4 min 4m 10s
Memphis
Guaymas
2 min 3 min 3m 30s 4 min
TOTAL SOLAR
Chihuahua
Ft. Worth Dallas
2 min 3 min 3m 30s
Shreveport
3 min 3m 10s 3m 40s 4m 20s 4m 20s
Jackson
4m 25s 4 min 4m 10s 3m 30s
2 min Austin
La Paz
4m 25s San Antonio Houston Baton Rouge
Culiacán
4m 25s 4m 20s 3 min
New Orleans
2 min 3 min 3m 30s 4m 20s Nuevo Laredo 2 min 4m 10s
Torreón Laredo Corpus Christi
Mazatlán 4 min 4m 25s 3m 30s
2 min 3 min
Monterrey Matamoros
2 min 3 min
San Luis Potosí

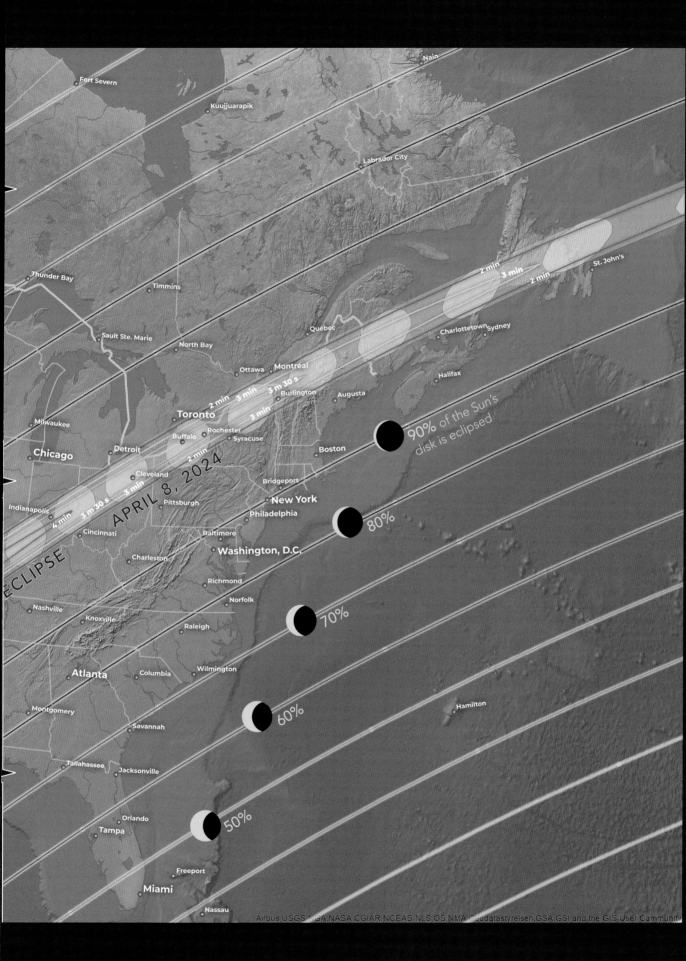

APRIL 8, 2024

ECLIPSE

90% of the Sun's disk is eclipsed

80%

70%

60%

50%

2 min   3 min
3 min
2 min   3 m 30 s
3 min
2 min
4 min   3 m 30 s   3 min

2 min   3 min
3 min   2 min

Nain

Fort Severn
Kuujjuarapik

Labrador City

Thunder Bay
Timmins
St. John's

Sault Ste. Marie
North Bay
Québec
Charlottetown Sydney

Milwaukee
Toronto
Ottawa Montréal
Halifax

Chicago
Detroit
Buffalo Rochester
Burlington Augusta
Cleveland
Syracuse
Boston

Indianapolis
Pittsburgh
Bridgeport

Cincinnati
Baltimore
New York
Philadelphia

Charleston
Washington, D.C.

Nashville
Richmond
Knoxville
Norfolk

Raleigh

Atlanta
Columbia
Wilmington
Hamilton
Montgomery

Savannah

Tallahassee
Jacksonville

Orlando
Tampa
50%

Freeport

Miami

Nassau

82%

84%

86%

88%

90%

92%

94%

96%

98% of the Sun's disk is occulted at maximum eclipse

BAJA CALIFORNIA SUR

La Paz

Cabo San Lucas

Ciudad Obregon

Navajoa

Huatabampo

El Fuerte

Los Mochis

Guasave

Guamúchil

SINALOA

Culiacán

Eldorado

La Cruz

Cuauhtémoc

Santa Barbara

DURANGO

12:15 pm MDT

Papasquiaro

Durango

12:10 pm MDT

Greatest eclipse at 12:05 pm MDT

2 min

3 min

3m 30s

4 min

4m 10s

4m 26s

Mazatlán

4m 27s

Escuinapa

3 min

2 min

4m 20s

4m 22s

4m 22s

4m 20s

4m 10s

4 min

3m 30s

Tecuala

4m 24s

Tuxpan

Valparaiso

NAYARIT

Tepic

Compostela

72

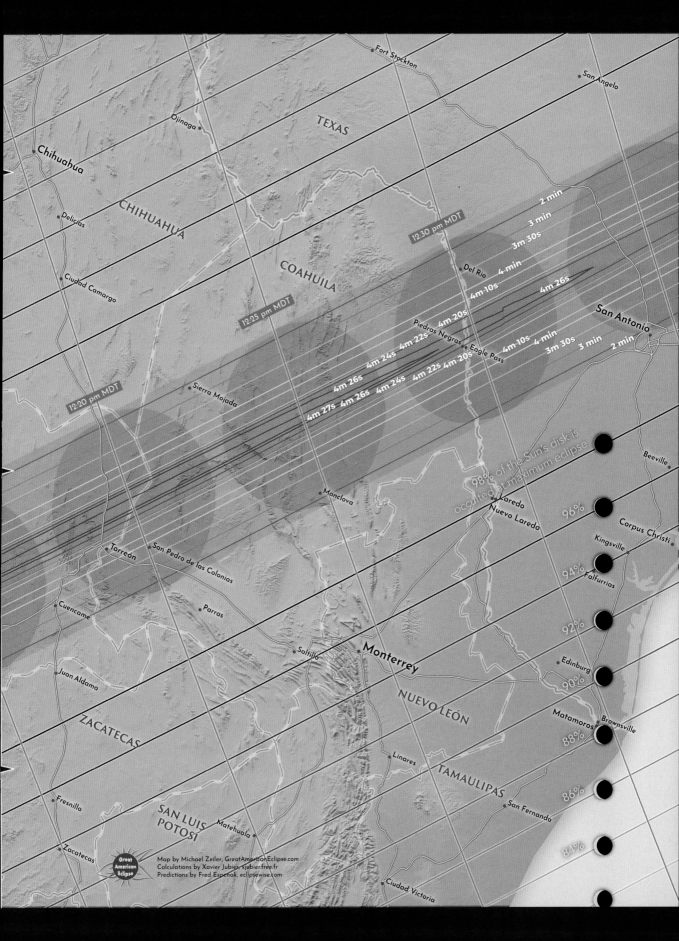

TEXAS

Fort Stockton

San Angelo

Ojinaga

Chihuahua

CHIHUAHUA

Delicias

Ciudad Camargo

COAHUILA

12:30 pm MDT

Del Rio

2 min

3 min

3m 30s

4 min

4m 10s

4m 26s

San Antonio

12:25 pm MDT

Piedras Negras   4m 20s

4m 22s

Eagle Pass   4m 10s   4 min

4m 24s   3m 30s   3 min   2 min

Sierra Mojada

4m 26s

4m 27s   4m 26s   4m 24s   4m 22s   4m 20s

12:20 pm MDT

Monclova

98% of the Sun's disk is
occulted at maximum eclipse

Beeville

Laredo

Nuevo Laredo

96%

Corpus Christi

Kingsville

Torreón

San Pedro de las Colonias

94%   Falfurrias

Cuencame

Parras

92%

Juan Aldama

Saltillo   Monterrey

Edinburg

90%

NUEVO LEÓN

Matamoros   Brownsville

ZACATECAS

88%

Linares   TAMAULIPAS

86%

Fresnillo

SAN LUIS
POTOSÍ

Matehuala

San Fernando

84%

Zacatecas

Great
American
Eclipse

Map by Michael Zeiler, GreatAmericanEclipse.com
Calculations by Xavier Jubier, xjubier.free.fr
Predictions by Fred Espenak, eclipsewise.com

Ciudad Victoria

Amarillo

Clovis

82%

Roswell

84%

86%

Carlsbad

Hobbs

Lubbock

OKLAHOMA CITY

Shawnee

Norman

Lawton

88%

TEXAS

Wichita Falls

90%

Midland

Pecos

Odessa

Sherman

Denton

Abilene

92%

Fort Stockton

San Angelo

Ft. Worth   Dallas

96%

4m 22s

98% of the Sun's disk is occulted at maximum eclipse

Waco

4m 24s

Killeen Temple

Del Rio 4 min   4m 10s   4m 20s   4m 22s

1:40 pm CDT

2 min   3 min   3m 30s   4m 26s   4m 22s   4m 24s

4m 20s

Austin

Huntsville

San Antonio

Piedras Negras   Eagle Pass   4m 10s

Conroe

4m 27s   4 min

1:35 pm CDT

3m 30s

TEXAS

Houston

3 min

Greatest eclipse
at 1:30 pm CDT

2 min

Victoria   Bay City

Monclova

Nuevo Laredo

Beeville

Port Lavaca

Freeport

Galveston

Laredo

Great
American
Eclipse

Corpus Christi

Kingsville

Map by Michael Zeiler, GreatAmericanEclipse.com
Calculations by Xavier Jubier, xjubier.free.fr
Predictions by Fred Espenak, eclipsewise.com

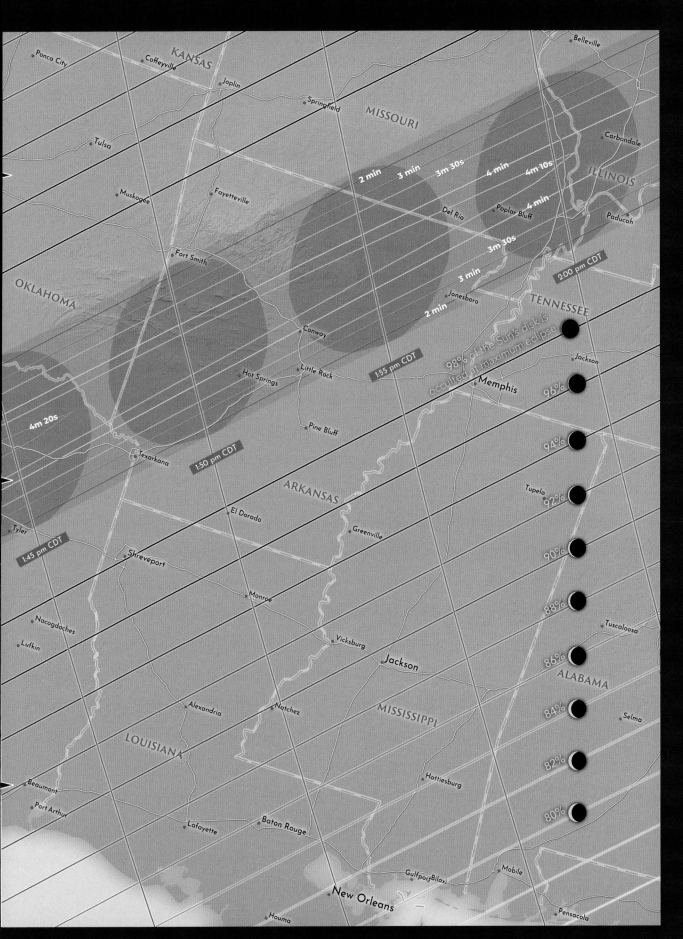

Ponca City
Coffeyville
KANSAS
Belleville

Joplin
Springfield
MISSOURI
Carbondale

Tulsa

2 min   3 min   3m 30s   4 min   4m 10s   ILLINOIS

Muskogee
Fayetteville
Del Rio   Poplar Bluff   4 min   Paducah

Fort Smith
3m 30s

OKLAHOMA
3 min

2 min   Jonesboro   TENNESSEE

Conway
98% of the Sun's disk is
occulted at maximum eclipse
Jackson

Hot Springs   Little Rock   1:55 pm CDT   Memphis   96%

Pine Bluff   94%

4m 20s
Texarkana   1:50 pm CDT

ARKANSAS   92%
El Dorado

Tyler   Greenville   90%

1:45 pm CDT
Shreveport   88%
Nacogdoches
Tuscaloosa

Lufkin
Monroe   86%

Vicksburg   ALABAMA
Jackson

Alexandria   Natchez   84%   Selma
MISSISSIPPI

LOUISIANA   82%

Hattiesburg
Beaumont   80%
Port Arthur

Lafayette   Baton Rouge

GulfportBiloxi   Mobile
New Orleans

Houma   Pensacola

75

84%

86%

Oshkosh

Fond du Lac

Cadillac

WISCONSIN
West Bend

88%
Madison

Milwaukee

Muskegon

Saginaw

Dubuque

90%
Racine

Grand Rapids

Flint

Waukegan
Lansing

Pontiac

92%
Aurora

Chicago

Kalamazoo

Ann Arbor

Detroit

Windsor

94%
Gary

MICHIGAN

96%
Kankakee

Toledo

3m 50s

Peoria

98% of the Sun's disk is occulted at maximum eclipse

INDIANA

Fort Wayne

Cleveland

Bloomington

Lima

Akron

Youngstown

Springfield

Decatur

Urbana

Kokomo

Canton

ILLINOIS

Muncie

3:15 pm EDT

3m 50s

2 min
3 min 3m 15s 3m 30s 3m 40s Indianapolis 4 min

Columbus

Wheeling

3m 50s

Bloomington

Dayton

OHIO

3m 40s

Cincinnati

Morgantown

3m 30s
Covington

3:10 pm EDT

3m 15s

3 min

Evansville

Louisville

Frankfort

WEST VIRGINIA

2 min

Owensboro

Greatest eclipse at 3:05 pm EDT

Lexington

Huntington

Charleston

Paducah

KENTUCKY

Bowling Green

Beckley

Clarksville

Nashville

TENNESSEE

Roanoke

82%

84%

Mistassini

86%

88%

90%

Dolbeau

92%

Chicoutimi

94%

Rimouski

QUEBEC

96%

98% of the Sun's disk is
occulted at maximum eclipse

Sept-Îles

Mingan

Gaspé

NEW BRUNSWICK

Bathurst

2 min   2m 30s 2m 40s 2m 50s   3 min

Edmundston

Québec

Shawinigan

Trois-Rivières

Victoriaville   Saint-Georges

2 min   2m 30s 2m 40s 2m 50s 3 min   3m 10s

3m 20s

3m 25s

3m 20s

3m 10s

3 min

2m 50s

3m 30s

2m 40s

2m 30s

2 min

Burlington

Presque Isle

Fredericton

3m 20s

PRINCE
EDWARD
ISLAND

3 min

2m 50s

2m 40s

2m 30s

2 min

Charlottetown

Moncton

3:35 pm EDT

MAINE

Bangor

Saint John

Windsor

Greatest eclipse
at 3:30 pm EDT

Augusta

Bar Harbor

Digby

Halifax

NOVA SCOTIA

Lewiston

Portland

Yarmouth

Shelburne

Concord

Map by Michael Zeiler, GreatAmericanEclipse.com
Calculations by Xavier Jubier, xjubier.free.fr
Predictions by Fred Espenak, eclipsewise.com

Great
American
Eclipse

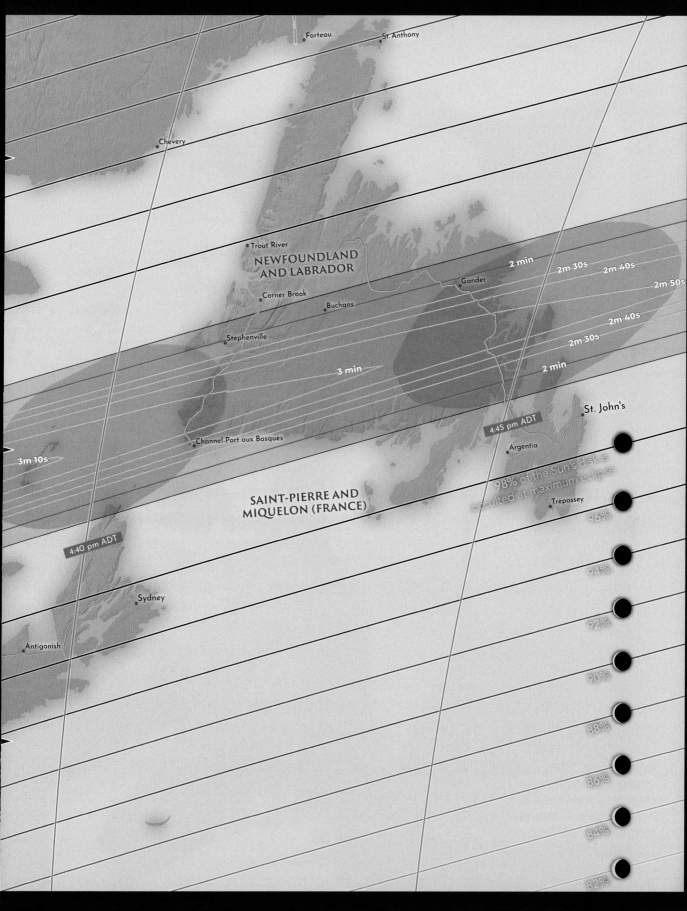

Forteau

St. Anthony

Chevery

Trout River

NEWFOUNDLAND
AND LABRADOR

Corner Brook

Buchans

Gander

2 min

2m 30s

2m 40s

2m 50s

Stephenville

2m 40s

2m 30s

3 min

2 min

Channel-Port aux Basques

St. John's

4:45 pm ADT

Argentia

3m 10s

SAINT-PIERRE AND
MIQUELON (FRANCE)

98% of the Sun's disk is
occulted at maximum eclipse

Trepassey

96%

4:40 pm ADT

94%

Sydney

92%

90%

Antigonish

88%

86%

84%

82%

Solar filters fitted to these binoculars make them safe for viewing the Sun during the partial phases of the eclipse.

© VASILIS VERVERIDIS | DREAMSTIME.COM

CHOOSING BINOCULARS FOR THE ECLIPSE

Some amateur astronomers consider binoculars an optional accessory, but most regard them as a necessity. Indeed, binoculars are the first thing that a beginning amateur should purchase to view the sky. And nothing is better for enhancing your view of totality.

However, do not view any of the solar eclipse's partial phases through unfiltered binoculars. Only totality is safe to view with your unprotected eyes or through binoculars. Some manufacturers make approved solar filters that will fit over the front lenses of your binoculars. With those in place, you can view the Sun at any time.

In some ways, binoculars may be a better choice than a telescope for the eclipse. Binoculars have a wide field of view and provide right-side-up images, making the Sun easy to find. They also require no effort or expertise to set up. And, for most people, observing with two eyes open rather than one seems more natural and comfortable.

Also, binoculars are easier on your wallet. Unless you're considering image-stabilized models, binoculars offer a more affordable way to view the eclipse than a telescope.

Binoculars use prisms to bend incoming light several times before it reaches your eyes. Manufacturers have to position all the optical components correctly so that no light is lost. HOLLEY Y. BAKICH

Celestron's Oceana 7×50 binoculars have great optics but weigh 41 ounces, a bit heavy for some eclipse watchers. CELESTRON

THE NUMBERS

Every binocular has a two-number designation, such as 7×50. The first number (in this case, 7) is the magnification, or power. The second number (50) is the diameter in millimeters of each of the objective lenses.

This example, 7×50, is great for astronomy. A magnification of 7 is in the "medium" range, just high enough to bring out some detail in large astronomical objects. Too high a magnification will over-magnify the involuntary motions of your hands. That motion causes celestial objects to move around. High magnification also limits the field of view, making objects more difficult to find for beginners.

In our example, 50 millimeters is the aperture (size) of each of the front lenses. The larger this number, the more light the binoculars can collect, and that makes the target brighter. Binoculars with 50-millimeter front lenses collect more than twice as much light as those measuring 35 millimeters across. The disadvantages to larger front lenses are that the binoculars will be larger, heavier, and more expensive.

OPTICS

The components inside binoculars that bend the light so the image appears right-side-up are the prisms. Binocular prisms come in two basic designs: roof prisms and Porro prisms (capitalized because they're named for 19th-century Italian inventor, Ignazio Porro). Roof prisms are more lightweight and smaller, but Porro prisms are better because they're made of a higher-quality glass.

Most high-quality binoculars are multicoated on all optical surfaces. You'll see this referred to as "fully multicoated." In the 1950s, manufacturers developed coatings that reduced light loss and internal reflections.

The eyepieces of Celestron's 25×100 SkyMaster binoculars focus individually. CELESTRON

MECHANICS

Center-focus binoculars move both optical tubes simultaneously. Other models let you focus each tube independently. All else being equal, choose binoculars with individual focus. Center-focus units add a bit of mechanical complication. Also, center-focus binoculars must allow you to adjust one of the tubes because most people's eyes do not come to the same focus.

Individual-focus binoculars tend to be more rugged and weatherproof. In either case, once you

Celestron's 8×56 SkyMaster DX binoculars use Porro prisms to bend incoming light to the eyepieces.
CELESTRON

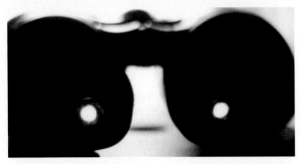

The round circles of light on the eyepiece lenses show the size of the binoculars' exit pupils. MICHAEL E. BAKICH

focus your binoculars on a sky object, the focus will be good for all other objects because all appear infinitely distant through binoculars.

EXIT PUPIL

One of the most important terms when dealing with binoculars is the exit pupil. This is the diameter of the shaft of light coming from each side of the binoculars to your eyes. If you point the front of the binoculars at a bright surface or the sky, you'll see two small disks of light at the eyepieces. These images show the lenses.

The diameter of the exit pupil equals the aperture divided by the magnification. So, for 7×50

Most binoculars designed for astronomy have their field of view printed directly on them. CELESTRON

binoculars, the exit pupil diameter would be 50 divided by 7, or roughly 7mm. The goal is to match the binoculars' exit pupil to your eyes' pupil size. By doing this, the image will be brighter because the light is hitting more of our eye's retina. If a binoculars' exit pupil is too large to fit into your eye, you will lose some of the instrument's incoming light.

You can measure your pupil size with a gauge available from some telescope suppliers. You also may be able to get one from a pharmaceutical company or from your optometrist.

FIELD OF VIEW

One number you'll usually find printed directly on the binoculars is the field of view, often abbreviated "FOV." High-quality binoculars (especially those designed for astronomy) give this number in degrees, and other models often state XXX feet at 1,000 yards. The latter lists the width of the field of view at a distance of 1,000 yards.

If your binoculars are the latter type, or if they contain no information on the field of view at all, don't worry. There's an easy way to figure out the approximate angular diameter (in degrees) of the field of view. Just start with the same math that you used to figure out the exit pupil: Divide the aperture (in millimeters) by the power. So, most 7×50 binoculars will have a field of view of 50/7 = 7.14°, or, what you'll more usually see and hear, 7°.

EYE RELIEF

Eye relief is a function of the eyepiece. It's the manufacturer's recommended distance (for best performance) your eye's pupil should be from the eyepiece's exit lens. Eye relief generally decreases as magnification increases. Eye relief less than 10mm requires you to position your eye quite close to the eyepiece.

Short eye relief poses no problem for advanced amateurs, but for beginners, longer eye relief

Celestron's Outland X 8×25 binoculars are a lighter choice, and weigh only 10.6 ounces. CELESTRON

allows the head more freedom of movement. Also, those who need to or choose to wear eyeglasses will require longer eye relief.

If you're given a choice, select binoculars that have an eye relief between 15 and 20 millimeters. If the eye relief is not listed, and if you wear glasses, be sure to try the binoculars with your glasses on before you buy them.

A BUYER'S GUIDE

How should you choose your binoculars for the eclipse? The best advice for many amateur astronomers is to conduct a short test before their purchase.

First, pick up the binoculars and shake them gently. Then twist them gently. Then move the focusing mechanisms several times. Then move the barrels together, then apart. What you're assessing is the quality of workmanship. If you hear loose parts or if there's any play when you twist or move the binoculars, don't buy them. Another thing to consider at this point is the weight of the binoculars. If you're going to be holding them in your hands, try to imagine what they will feel like at the end of the eclipse.

Canon's 18×50IS binoculars keep objects in view steady by compensating for involuntary motions of your hands.

MICHAEL E. BAKICH

Look into the front of the binoculars and check for dirt or other contaminants. The inside of the binoculars should be immaculate. If not, don't buy them.

Hold the binoculars in front of you with the eyepieces toward you. Point them at something bright. You'll see the exit pupils—disks of light formed by the eyepieces. They should be round. If they are not round, the optical alignment of the binoculars is bad, and the prisms are not imaging all the light.

You should look through the binoculars as well. Try to do this outdoors. If that's impossible, look through a door or window at distant objects. How well do the binoculars focus? Are objects clear? If there is any sign of a double image, the two barrels are not aligned. Don't buy them.

If you are wearing glasses—and if you plan to observe with them on—there are other questions. Can you get your eyes close enough to the binoculars to see the entire field of view? If possible, aim the binoculars at a straight line such as a phone wire or the horizon, if possible. Does the line look straight? A tiny amount of bending near the edges of the field of view is not a big problem. But if you see a lot, steer clear.

Repeat the above tests with several different binoculars. Once you become more familiar with how binoculars compare, you will be well on your way to purchasing an excellent unit.

IMAGE-STABILIZED BINOCULARS

To enjoy both a wide field of view and moderately high magnification in hand-held binoculars is to have the best of both worlds. Manufacturers have achieved this by creating image-stabilized (IS) binoculars.

The optics and mechanics of IS

This bracket is the simplest way to attach binoculars to a tripod. MICHAEL E. BAKICH

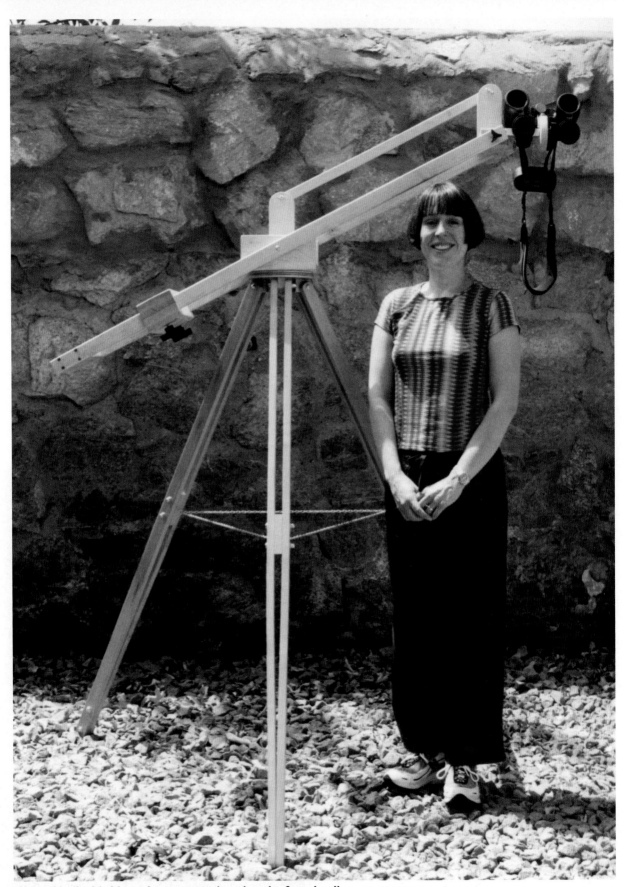

Michael built this binocular mount using plans he found online. MICHAEL E. BAKICH

Always place approved solar filters at the front of your binoculars. MICHAEL E. BAKICH

binoculars vary in the same way as the optics of regular binoculars, so you should apply the same tests I listed above if you are considering purchasing a unit. And remember, technology like this isn't free, or cheap. IS binoculars large enough to interest amateur astronomers cost, at this writing, between $1,000 and $2,000.

MAINTENANCE

Caring for your binoculars is easy. Most units come with lens caps, eyepiece caps, and a case. Use them. These protect your binoculars from dust and moisture. Don't leave your binoculars exposed to direct sunlight, even if they're in their case. Heat will cause the carefully placed elements of binoculars to expand and later contract.

Cleaning is only sensitive when it involves the lenses. If your lenses become dusty, blow them off with compressed air or brush them with an approved optics brush. You'll find both of these at any camera store. If you must wipe the lenses, use only lens paper, and change it frequently, rather than using the same piece to wipe back and forth.

The body of your binoculars also will get dirty. When it does, simply wipe it with a damp cloth. And because binoculars are aligned optical equipment,

keep the vibrations, especially impacts, to a minimum.

BINOCULAR MOUNTS

Image-stabilized binoculars are a tremendous advance for amateur astronomy. But for the steadiest images possible, nothing beats mounting your binoculars to a tripod or custom binocular mount. Smaller, well-mounted, low-power binoculars always will beat bigger hand-held binoculars.

The simplest binocular mount is a metal "L" bracket. Attach it to the ¼-20 mounting hole on the binoculars' center post. The other end of the L attaches to a camera tripod.

Another option is to purchase or build a custom-made binocular mount. A quick search of the Internet will uncover several plans for high-quality binocular mounts.

When selecting a binocular mount, choose one that is sturdier than you require. That way, you can easily upgrade to a larger model in the future. A binocular mount is rugged if, a few seconds after you have located an object, the image settles down

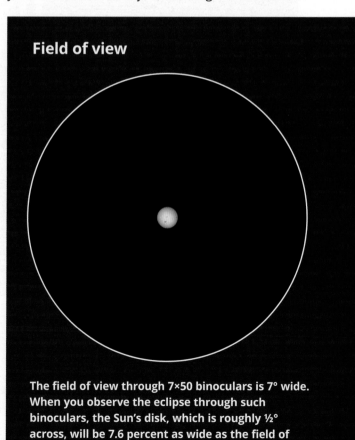

Field of view

The field of view through 7×50 binoculars is 7° wide. When you observe the eclipse through such binoculars, the Sun's disk, which is roughly ½° across, will be 7.6 percent as wide as the field of view. SOHO/NASA

and shows no vibration (unless a strong wind is blowing).

The other necessity for high-power binocular observing is the tripod the mount attaches to. Most camera tripods are not robust enough to handle the weight of the binoculars plus the weight of the mount. If you have a tripod, by all means, try it. You'll know immediately if it's up to the task.

SUGGESTIONS FOR THE ECLIPSE

Now that you know what goes into buying binoculars, which one should you choose? First, I can recommend six manufacturers: Canon (www.usa.canon.com), Celestron (www.celestron.com), Fujinon (www.fujifilmusa.com), Nikon (www.nikonusa.com), Orion (www.telescope.com), and Vixen (www.vixenoptics.com). These companies make high-quality binoculars, they don't make wild claims, and they stand behind their products.

There are two suggested ways for determining how large the Sun will look through your binoculars. 1) Observe the Full Moon. The disks of the Sun and the Moon are the same size. 2) Purchase approved solar filters for your binoculars and begin observing the Sun now.

These methods show only how the Sun's disk will appear through your binoculars. Remember, however, that during totality you also will see the corona, our star's thin outer atmosphere. During some eclipses the corona has stretched nearly three times the disk's diameter from the Sun in all directions, which would make it 3.5° across at maximum.

For most of the 14 eclipses I have observed, I've used 7×50 binoculars. But during the past three or four, I've gone with an 18×50 image-stabilized model. Those have a smaller field of view (3.7°), but they make the streamers in the corona a bit easier to see.

Of course, you'll see the main spectacle best using just your naked eyes. I will have binoculars around my neck for the eclipse. If I had to guess, I'd estimate that I'll view the roughly 4 minutes and 27 seconds of totality like this: 3 minutes with naked eyes; 87 seconds through binoculars. But as they say, your mileage may vary.

The field of view through 10×50 binoculars is 5° wide. When you observe the eclipse through such binoculars, the Sun's disk will be 10.6 percent as

The field of view through Canon's 18×50 image-stabilized binoculars is 3.7° wide. When you observe the eclipse through such binoculars, the Sun's disk

A telescope is a great way to see all of the detail of the Sun both during the partial and total phases of the eclipse. Be sure you have the proper solar filters for looking at the Sun at any time other than totality.

VASILIS VERVERIDIS/DREAMSTIME

CHOOSING A TELESCOPE FOR THE ECLIPSE

Buying a telescope is a big decision, and you shouldn't rush into it just because the total solar eclipse is coming. This chapter will get you started.

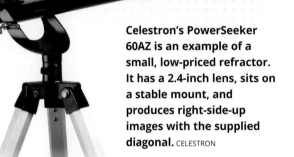

Celestron's PowerSeeker 60AZ is an example of a small, low-priced refractor. It has a 2.4-inch lens, sits on a stable mount, and produces right-side-up images with the supplied diagonal. CELESTRON

THE NUMBERS

The main number of any telescope is its aperture, the size of the main lens or mirror. Manufacturers give it in inches. Numbers like "f/4," "f/10," or "f/15" are a telescope's focal ratio. This number can give you the approximate length of any telescope. For example, a 6-inch f/4 telescope will be approximately 6 inches times 4, or 24 inches long (excluding lens shades, etc.). A 6-inch f/10 telescope, on the other hand, will be roughly 60 inches long.

REFRACTING TELESCOPES

Refraction is the bending of light that happens when it passes from air to glass and back. A refract-ing telescope, or refractor, uses carefully made lenses to bring the light from objects to a focus.

The earliest telescopes had poor optical quality because the lenses had various defects. In 1729, the first lens that combined two different types of glass appeared. The word for this type of lens is "achro-mat." An achromatic lens does a good job of bringing all colors of light to the same focus.

In the 1980s, the first "apochromatic" lenses became available. An achromat is a two-lens system. Apochromats also may use two lenses, but they're more likely to have three or four. The main differ-ence between the two types is the amount of color fringes you'll see on bright objects. It usually appears

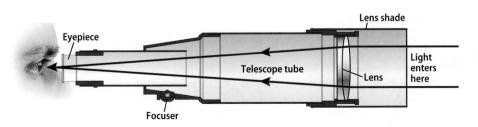

Eyepiece

Telescope tube

Focuser

Lens shade

Light enters here

Lens

A refractor uses a lens (a combination of two to four glass pieces) to bring light to a focus. ASTRONOMY: ROEN KELLY AFTER CELESTRON

as a purple fringe on one edge of a bright object like Jupiter, Venus, or the Moon. The image in an apochromat has no fringes. Because of the additional lens elements, however, apochromats aren't cheap.

REFRACTORS: PLUSES AND MINUSES

Lenses don't have a central obstruction. That means image contrast is generally better in refractors.

The second advantage of refractors is that they are low maintenance. Lenses never require recoating as mirrors do. Also, a lens system generally doesn't get out of alignment unless the scope encounters some major trauma. In other words, if you don't drop the telescope, you'll never have to align it.

Yet another advantage is that some of the newest refractors are among the smallest telescopes made. That makes them the most portable.

The only downside is that because a refractor has a closed tube, it requires a certain amount of time to adjust to the outside temperature. Today's thin-walled aluminum tubes have reduced this period significantly, but you still have to take it into account.

REFLECTING TELESCOPES

Scottish mathematician James Gregory invented the reflecting telescope and published a description in 1663. But he never actually produced a telescope. English mathematician Sir Isaac Newton constructed the first working reflector telescope in 1668. It had a mirror 1 inch across and a tube 6 inches long.

A Newtonian reflector contains two mirrors — one called the "primary" at the bottom of the tube and a small, flat "secondary" near the top of the tube. Light enters the top, travels down the tube, hits the primary, and reflects to the secondary mirror. That mirror then reflects it into the eyepiece.

REFLECTORS: PLUSES AND MINUSES

Unlike refractors, reflectors show no color fringes around even the brightest objects. Their biggest

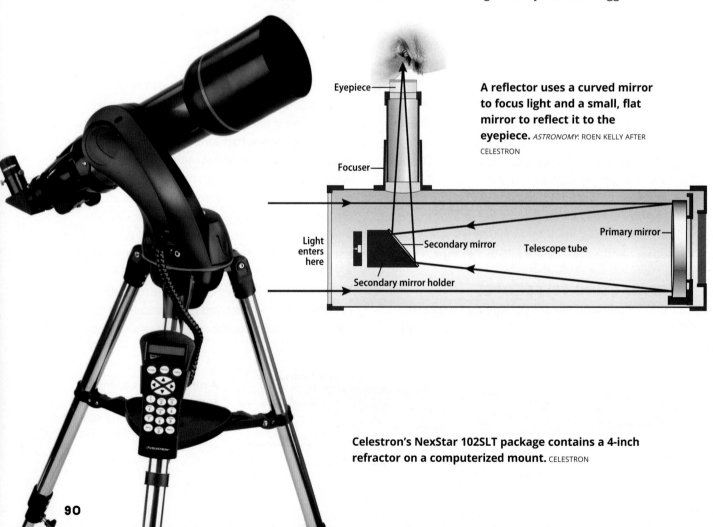

Eyepiece

A reflector uses a curved mirror to focus light and a small, flat mirror to reflect it to the eyepiece. ASTRONOMY: ROEN KELLY AFTER CELESTRON

Focuser

Light enters here

Secondary mirror

Primary mirror

Telescope tube

Secondary mirror holder

Celestron's NexStar 102SLT package contains a 4-inch refractor on a computerized mount. CELESTRON

advantage, however, is cost. When working with a mirror, manufacturers have to polish only one surface. An apochromatic lens has between four and eight surfaces, plus you're looking through the lenses, so the glass has to be defect-free. Telescopes with apertures of more than 6 inches, with few exceptions, are all reflectors or compound telescopes.

One disadvantage is that the secondary mirror is an obstruction that scatters a tiny amount of light from bright areas into darker ones. Most people never notice this.

Newtonian reflectors suffer from "coma," a defect that causes stars at the edge of the field of view to look like a comet. Some manufacturers

Celestron's AstroMaster 130EQ is a 5.1-inch reflector on an equatorial mount. CELESTRON

Celestron's SkyProdigy 130 is a reflector that contains a 5.1-inch primary mirror. It sits on a computerized alt-azimuth mount. CELESTRON

Primary mirror

Telescope tube

Light enters here

Eyepiece

Secondary mirror

Focus knob

Corrector plate

A compound telescope combines a front lens with mirrors to focus light. This diagram shows a Schmidt-Cassegrain telescope. *ASTRONOMY*: ROEN KELLY AFTER CELESTRON

that have a mix of refractor and reflector elements in their design. German astronomer Bernhard Schmidt made the first one in 1930. The Schmidt telescope had a spherical primary mirror at the back of the instrument and a glass corrector plate in the front.

The Schmidt telescope was the precursor of today's most popular design, the Schmidt-Cassegrain telescope, or SCT. In this design, light enters the tube through a corrector plate. It then hits the primary mirror at the tube's base, which reflects the light to a secondary mirror mounted on the corrector. The secondary reflects light through a hole in the primary mirror to the eyepiece, which sits at the back of the scope.

CATADIOPTRICS: PLUSES AND MINUSES
The number one advantage of an SCT is its compact design. Such instruments are often only one-quar-

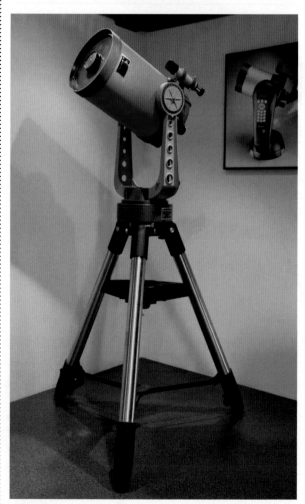

Celestron's original C8 had an orange tube. CELESTRON

Celestron's NexStar 127SLT is a 5-inch compound telescope supplied with a go-to mount that runs on eight AA batteries (or an optional adapter). CELESTRON

make correcting lenses that fix this problem. Observers usually just keep all targets at the center of the field of view.

Finally, because of how the mirror attaches to the tube, a reflector is sensitive to bumps when transported. Usually, reflectors have to be aligned before each observing session.

CATADIOPTRIC TELESCOPES
The third type of telescope, the catadioptric, is also known as a compound telescope. They are hybrids

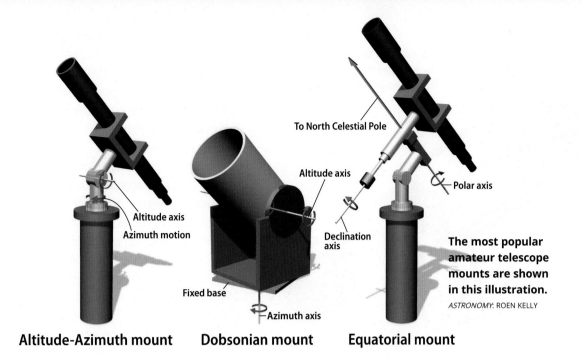

To North Celestial Pole

Altitude axis

Altitude axis

Azimuth motion

Declination axis

Polar axis

Fixed base

Azimuth axis

The most popular amateur telescope mounts are shown in this illustration. *ASTRONOMY*: ROEN KELLY

Altitude-Azimuth mount   **Dobsonian mount**   **Equatorial mount**

ter as long as comparably sized reflectors and much shorter than refractors with half their aperture. This feature makes the SCT one of the ultimate grab-and-go telescopes.

However, like refractors, compound telescopes also have closed tubes. Adjusting to the outside temperature, therefore, takes a while.

MOUNTS AND DRIVES

We call our instruments telescopes, but the phrase "optical tube assembly on a mount" also works. In fact, it points out that half of any telescope is the mount. Some observers claim that the mount is the more important part. An unstable mount will render the finest telescope unable to deliver quality images.

An alt-azimuth mount is the simplest type of telescope mount. The name is a combination of the words "altitude" and "azimuth." Altitude is the number of degrees a sky object lies above the horizon. Azimuth is the number of degrees from north to the object moving around the horizon. A telescope on this type of mount moves up and down (altitude), and left and right (azimuth). A camera tripod is a simple alt-azimuth mount.

Telescope maker John Dobson invented a simple type of alt-azimuth mount that now bears his name. The Dobsonian mount is the least expensive telescope mount and manufacturers almost always

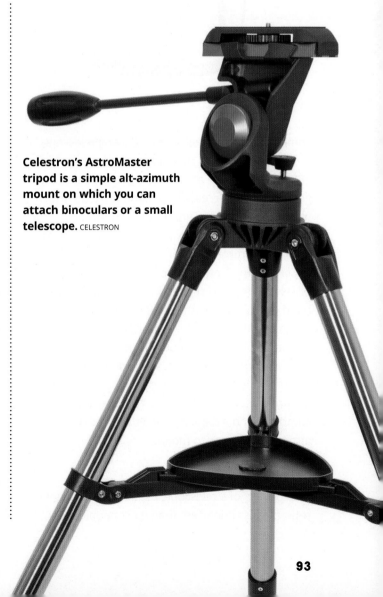

Celestron's AstroMaster tripod is a simple alt-azimuth mount on which you can attach binoculars or a small telescope. CELESTRON

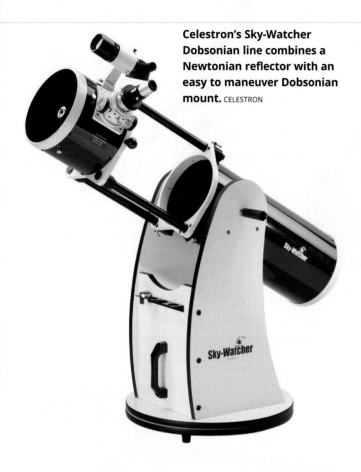

Celestron's Sky-Watcher Dobsonian line combines a Newtonian reflector with an easy to maneuver Dobsonian mount. CELESTRON

Claims of 500x for small telescopes are meaningless, and here's why: All you need to do to change the magnification of a telescope is change the eyepiece. So, if a low-quality, high-power eyepiece comes with the telescope, the instrument can achieve high magnifications. But it won't be worth looking through.

To calculate the magnification of any eyepiece, just divide the telescope's focal length (listed in the instruction manual, on the tube, or on the front lens of some refractors and SCTs) by the eyepiece's focal length (the number printed on the eyepiece). For example, let's say that a certain telescope has a focal length of 1,000 millimeters. If you choose a 25mm eyepiece, the magnification will be 1,000 divided by 25, or 40x. If you replace the 25mm eyepiece with a 10mm eyepiece, the magnification will be 1,000 divided by 10, or 100x.

EYEPIECES

The best eyepieces contain multiple highly polished and coated lenses made from exotic glass, so they

combine it with a Newtonian reflecting telescope.

A development in mounts that occurred in the 1980s is the driven alt-azimuth mount, also called a go-to mount. To create this, manufacturers attach motors to both the altitude and azimuth axes. The motors also interface to a computer. Once you run through a simple setup procedure, you're ready to observe. Simply enter the object's name or number, or select it from a list contained in the telescope's hand controller, and the go-to drive will find your target and then continues to track it.

The second type of mount is the equatorial mount. German optician Joseph von Fraunhofer invented it in the early 19th century to track the apparent motion of the stars. With the addition of a computer and a database, an equatorial mount also becomes a go-to mount.

POWER!

The "power" of a telescope, usually called magnification, is less important than most people think.

Celestron's NexStar 6SE package combines a 6-inch Schmidt-Cassegrain telescope, a tripod, and a computerized go-to mount. CELESTRON

Celestron's Eyepiece and Filter Kit contains five eyepieces, a Barlow lens, six color filters, and a Moon filter. CELESTRON

are not cheap. Coatings, by the way, are ultra-thin layers manufacturers apply to lenses to reduce the amount of light reflected by the eyepiece's glass and thus increase the amount that passes through them. You won't need top-of-the-line eyepieces if your telescope is just for the eclipse.

When choosing which eyepiece to buy, consider its weight. Some tip the scale at more than 2 pounds—as much as some binoculars. If you purchase a small or medium-sized telescope, you'll want to choose lighter eyepieces.

Another thing to keep in mind is the eyepiece's field of view. You'll see two numbers used, the apparent field of view and the true field of view. The apparent field of view of an eyepiece is the angle of light able to enter the eyepiece. Eyepiece apparent fields range from 25° to 84°. An eyepiece's true field

is the amount of sky you actually see when you look through the eyepiece. This number can change from one telescope to the next.

SUGGESTIONS FOR THE ECLIPSE

One thing you can do to get ready is to start observing the Sun now. To do that, and to be able to safely view the partial phases of the eclipse, you'll need an approved solar filter. A filter that fits over the front of your telescope is the only kind to use. Never use a filter that screws into an eyepiece to view the Sun. Be absolutely certain wind or accidental bumps cannot dislodge your filter. If you are in doubt, securely tape the side of the filter to the telescope tube.

Start your solar observing by counting sunspots. Apart from being fun, sunspot counts let you know

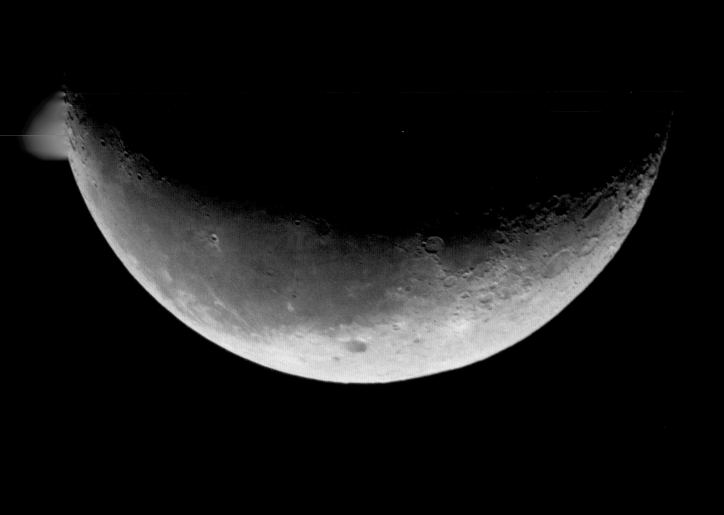

To practice for the April 8, 2024, total solar eclipse, observe the Moon through the same setup you'll use

how active the Sun is. And if you watch from day to day, you'll be able to see that the Sun rotates. People have been recording sunspot numbers on a daily basis since the middle of the 17th century.

If you don't have a solar filter yet, you can observe the Moon. Viewing Earth's satellite will familiarize you with how a Sun-sized object looks through your telescope through different eye-pieces.

Next is the selection of eyepieces. Be sure when you select a telescope and an eyepiece that the combination produces a true field of view (TFOV) wide enough for the eclipse. The Sun and Moon are each approximately ½° wide. Obviously then, don't choose a telescope-eyepiece combo whose TFOV is less than that.

When I've used a telescope to view solar eclipses, I carefully select just two eyepieces to combine with it. The first one, which has higher magnification, gives a TFOV of 1°. That makes the Sun's disk half as wide as the whole field of view. The second eye-piece provides a TFOV of 2°, which makes the Sun one-quarter the diameter of the field.

The first eyepiece is for the partial phases of the eclipse. It makes the Sun large enough to see any sunspots. As totality approaches, however, I switch to the lower power eyepiece because I want to see the Sun's corona, which appears as a ring around the eclipsed disk. Simple math shows that the ring of sky around the Sun you'll see through the higher-magnification eyepiece will be ¼° wide. The lower-power eyepiece, however, will produce a ring around the Sun that's ¾° wide. If I want a wider view yet, I switch to binoculars.

You can calculate an eyepiece's TFOV if you divide the eyepiece's apparent field of view by its magnification when it's in your chosen telescope. To get this result, first calculate the magnification, which I discussed above. Then divide that number into the eyepiece's apparent field of view. Let's look at an example.

Your scope has a focal length of 500mm, and you want the TFOV with your 25mm eyepiece, which has an apparent field of view of 50°:

Magnification = focal length of telescope / focal length of eyepiece

Magnification = 500 / 25 = 20x

If you attach a finder scope to your telescope, make sure it also has an approved solar filter. If not, remove the finder scope. CELESTRON

TFOV = apparent field of view / magnification

TFOV = 50° / 20 = 2.5°

For this telescope/eyepiece combo, your TFOV would be 2.5°. That would make the diameter of the Sun's disk 20 percent as wide as the view through the eyepiece, and it would leave a 1° ring around it.

How about a 10mm for your second eyepiece? If it also has an apparent field of view of 50°

Magnification = 500 / 10 = 50x

TFOV = 50° / 50 = 1°

This would be an excellent pair of eyepieces for the telescope described above.

Allow me to share one more piece of advice. Keep things simple. Don't bring more than two eyepieces for eclipse viewing. This will make the transition from your high-power to low-power views foolproof. You won't grab the wrong eyepiece or be looking down to be certain you get the right one. Remember: Time is of the essence!

IMPORTANT ECLIPSE-RELATED NOTE: If you will be using a telescope to observe the eclipse, either remove your finder scope, cover it, or purchase an approved solar filter for it, and secure it so it won't fall off (or be easily removed). That way, neither you nor anyone else will mistakenly look through the finder at the Sun's full glory. To say that would be bad is an understatement.

A camera with a solar filter was a popular attraction for people who wanted to observe the partial solar eclipse in Vancouver, British Columbia on August 21, 2017.
MODFOS/DREAMSTIME

CHOOSING A CAMERA FOR THE ECLIPSE

I was *Astronomy*'s photo editor for nearly two decades, so I can help you figure out what features to look for if you'll be purchasing a camera for the eclipse. But first, let me save many of you some time. For best results, you must shoot through a digital single-lens reflex camera (DSLR). If you plan to photograph the eclipse with your phone or with a point-and-shoot camera, you can stop reading this chapter now. Those devices will allow you to capture a memory, but little else.

WHICH SENSOR?

Having a larger sensor (often called the camera's chip) is one of the main reasons you'll get much better results when you shoot the eclipse through a DSLR. Even with a DSLR, however, you have two choices when it comes to the sensor inside. The smaller one is the Advanced Photo System type-C (APS-C) sensor.

The second choice is a full-frame sensor, which measures 36mm by 24mm, the same size as 35mm film. Full-frame sensors collect more light than their smaller counterparts. This translates to a better overall image.

The larger sensors also capture more of an image, while smaller ones crop in on the image,

This graphic shows the ratios of chip sizes in various cameras. The blue rectangle represents a full-frame DSLR chip. The yellow shows the APS-C sensor. The red is the average size of the sensor in a point-and-shoot camera, and the black rectangle shows the size of a cellphone chip. HOLLEY Y. BAKICH

producing a view with a narrower angle. For example, let's say you take a picture through a full-frame DSLR with a 20mm lens. If you took the same shot using the same lens through a camera with an APS-C sensor, the view would appear like you used a 31mm lens on the full-frame DSLR. One benefit, then, of the larger sensor is that it makes wide-angle photography easier.

The other main difference between DSLRs with the two types of sensors is that full-frame models are more expensive. When we consider the upcoming solar eclipse, is the extra expenditure worth it? No, it's not, in my opinion. All things being equal, you will get images from APS-C sensors that will blow your mind. Why spend $2,000 or more for a camera (just the body, now, we're not adding in lenses yet) when a $500 model will do?

Canon's EOS 7D is an inexpensive camera body that contains an 18-megapixel chip. COYAU/WIKIMEDIA COMMONS

MORE MEGAPIXELS?

Manufacturers measure camera resolution in megapixels, which equals 1 million pixels. Each pixel is a light-collecting element on the camera's sensor. What then makes a camera a 12-megapixel model? If its chip contains rows 4,000 pixels wide and columns 3,000 pixels high, then 4,000 × 3,000 = 12,000,000 pixels, or 12 megapixels.

Because of all the advertising on television and in magazines, you probably think that a sensor with more megapixels will deliver better resolution, and, therefore, a superior image. That would be true only if all pixels were the same. They're not.

Ads for cellphones often tout huge numbers of megapixels. 40 million pixels on a sensor? Wow! But wait a minute. Cellphone chips are roughly 2 percent the size of a full-frame sensor. To pack those millions of pixels onto such a small sensor, camera phones use much smaller pixels than DSLRs. The larger pixels used on full-frame or APS-C sensors capture more light, produce better color, and have less noise.

THE CAMERA'S PROCESSOR

If you're a beginning photographer, or even one with some experience who will be upgrading your camera, having a high-quality image processor will help you a lot. If you're more experienced, and especially if you shoot RAW images rather than

JPEGs, the internal processor won't matter much to you. RAW images are the unprocessed data the sensor sees. This setting is for those who want to process the images using software like Photoshop.

The reason most people shoot JPEGs rather than RAW images is that JPEG is a compressed format. So, it takes up less space on your camera's memory chip. If you shoot JPEGs, the processor matters. Processors that produce JPEGs handle operations that can fix lighting issues and adjust various other settings. It also lets the camera capture images in quick succession. While this doesn't affect image quality directly, being able to capture images quickly can mean the difference between a good shot and a great shot.

LENS DIFFERENCES

The camera's lens is its eye to the universe. If it doesn't perform well, you will not be happy with your images. Cellphones and point-and-shoot cameras won't produce high-quality images because their lenses are tiny and fixed to the camera. Manufacturers measure lenses by their aperture, which is how wide the lens can open. A wider aperture means more light. More light means you can take photos more easily where there isn't a lot of it.

But you may think because we're dealing with the Sun that there will be plenty of light. Not during totality. The illumination during the total phase of the eclipse will be only one-millionth what it was when the disk was visible. So, indeed, you do need to factor in a lens' light-grasp ability.

ECLIPSE CONSIDERATIONS

Before we consider which DSLR to buy, let me tell you where to look: eBay. I dialed up the giant online auction house, entered a few criteria, and was amazed at the quality of camera bodies you can get for less than $500. About 50 percent were new, and the others were "like new." And I was just surfing the Buy It Now options. You might be able to get a better deal if you select Auction.

I selected "Canon" as my manufacturer, but "Nikon" will work equally well. I'm more familiar with Canon cameras. More than 90 percent of the images I received at *Astronomy* magazine originated in that brand. But Nikon also produces high-quality gear.

Set 12 megapixels as your minimum chip criteria for imaging. That will produce images with dimen-

sions of 4000×3000 pixels, which will allow you to zoom in for your final composition if that's what you want.

Finally, if you will be shooting through a telescope, remember that you'll need a T-adapter and a T mount. See item no. 13 in Chapter 12 for more about this.

A telephoto lens attached to Canon's EOS 6D would provide great images of the eclipse. RICHARD BARTZ/WIKIMEDIA COMMONS

You can expect crowds in photogenic spaces on eclipse day. Getting your gear set up early may be essential to get the shots you want. DAVID CAREY/DREAMSTIME

A no. 14 welder's glass is a safe solar filter. You can view the Sun through it any time. Tests have shown no. 12 welder's glass is also safe, but most people find the Sun's image through it too bright. MICHAEL E. BAKICH

THE RIGHT FILTERS FOR THE ECLIPSE

We know that viewing the Sun can be dangerous, but it can be done—even through binoculars and telescopes—if you have the proper equipment. I categorize such devices into two categories: handheld (which include welder's glass and solar glasses) and optic-mounted filters. And while I called the first category "handheld," it includes both welder's filters and solar viewing glasses.

WELDER'S GLASS

Before about 1930, the original handheld solar filter was a piece of smoked glass. This gave a rather poor image and if the surface wasn't evenly covered by the soot, viewing through it could be unsafe. In the early 1930s, welder's goggles came into wide use and, in 1937, the welding helmet was invented by Willson Products of Reading, Pa.

To observe the Sun safely, you must use a no. 14 welder's filter. Unfortunately, welder supply stores generally do not stock no. 14 filters because they're the darkest ever made. You can still order them from a shop or online at a cost of only a few dollars each.

Most observers don't like them because they impart a greenish hue to the Sun. That color doesn't bother me. I have been using the same 4¼ x 2-inch piece of no. 14 welder's glass since my first total solar eclipse in 1970.

You might still find a welder's glass that's glass. Some manufacturers have started making them out of polycarbonate material. Experts initially thought them unsafe for solar viewing until June 2015, when a new standard emerged as the international guideline for safe solar eclipse viewing.

One of its architects, B. Ralph Chou, Professor Emeritus, School of Optometry and Vision Science, University of Waterloo, said, "The essential part of the standard is that solar filters are to have luminous transmittance equivalent to between Shade Number 12 and 16, and there are additional constraints on UV and IR transmittance, but these are not the same as the transmittance requirements for welding filters. In our rationale for the transmittance levels, David Sliney and I wrote that the UV and IR levels in sunlight are not a significant factor for solar retinopathy—the injury is primarily due to short-wavelength visible light."

This means it's OK to use a polycarbonate welder's filter if you can find one with a dark enough shade (no. 14), though some differences do exist between them and the glass versions.

As for differences, glass lenses can break. Polycarbonate lenses will never break. Polycarbonate lenses scratch much easier than glass ones, so glass lenses will look much better over time, provided you don't drop them. Finally, glass welding lenses have reasonably good optical quality (as good as a ¼-inch-thick piece of glass can have); polycarbonate's optical quality is significantly lower.

SOLAR GLASSES

Today, the easiest and cheapest way to view the Sun is through solar viewing glasses. These devices have several advantages over welder's glass. Eclipse glasses are made of cardboard that hold filters made of solar Mylar—which is silver—or a black polymer (plastic) and are much lighter. The advantage of the polymer is that it holds up to use (and unintentional abuse) better than Mylar.

You don't have to hold eclipse glasses to view the Sun—you wear them. And although you can slip your welder's glass into goggles designed to hold them, that combination is heavy and bulky. And if you drop each of these on a hard surface, the glasses will survive.

One of the two large suppliers, American Paper Optics, claims to have sold two billion eclipse glasses in the past quarter-century. The retail price for these glasses is in the $2 range, less for quantities.

The other main retailer, Rainbow Symphony, recently introduced a line of designer eclipse glasses. Rather than cardboard, the new models have frames of molded plastic. Some were standard glasses, and others used a wrap-around style. All are safe for solar viewing. In fact, you can't see anything but the Sun through these glasses.

The one possible issue with designer glasses is

Rainbow Symphony's solar viewing glasses may resemble standard sunglasses, but the only thing you can see through them is the Sun. MICHAEL E. BAKICH

that someone might think you're watching the Sun through sunglasses, which nobody should ever do. If anyone's around when you have such glasses on, make sure to tell her or him about safe solar viewing. You can even let them try on the glasses for their first safe look at the Sun.

TELESCOPE FILTERS

The most important thing about using a filter with a telescope is that the filter always goes on the front end of the scope. Never use any filter that fits over or screws into an eyepiece. They have been known to crack, and if you're looking through one when that happens, eye damage will happen.

The oldest option uses a piece of flat, polished glass coated with aluminum, nickel, or chromium to

Michael constructed this filter by combining a hardwood frame and Baader AstroSolar Safety Film. The unit fits snugly over the front end of a telescope, and the black knob at the top, attached to a threaded brass rod, secures it. MICHAEL E. BAKICH

The Solarscope
MICHAEL E. BAKICH

drop the Sun's brightness to safe and comfortable levels. Most glass filters impart an orange color to the image.

Another type of Sun filter for your telescope is made of solar Mylar mounted in a metal or plastic cell. Usually, these cost less than glass filters. Mylar filters make the Sun appear slightly blue.

The third type of filter uses the "solar safety film" developed by Baader Planetarium in Germany. The filter material is a high-strength polymer metalized on both sides. Baader astrofilm provides a white image of the Sun.

Don't be concerned if you see wrinkles in the Mylar or Baader filters. Because your telescope isn't focusing on them, the image quality isn't affected. A "stretched flat" film filter is more likely to tear from even small impacts.

Finally, companies like DayStar Filters and Rainbow Symphony offer lines of polymer filters designed to fit the front ends of binoculars, telescopes, camera lenses, and even finder scopes.

All of these filters work well, and they come in two different types: full-aperture and off-axis. The first type covers the whole front of the scope. You'll use one of these if your telescope is a refractor. The second provides a smaller opening away from the center of reflectors and catadioptric telescopes,

where the secondary mirrors sit.

Other options to welder and solar viewing glasses are available. Some are filters you add to existing telescopes, and some are stand-alone solar scopes. I currently own many of them.

SOLARSCOPE

Using no filters at all, the Solarscope projects the Sun's image. Set up this simple solar viewer during the partial phases of the eclipse, and you're sure to attract a crowd. It's an ideal product for Sun-viewing that combines safety and low cost.

The Solarscope was invented by astronomer Jean Gay from Cote d'Azur Observatory in Nice, France, as an easy way for groups to observe the Sun. The product produces an image that's totally safe to view.

The viewer comes in two sizes, for either one person or a group, and is made of sturdy cardboard. You have to put it together, but the manual has simple directions. With the Solarscope's cleverly designed base and hood assembly, the Sun's image projects onto a dark area, which makes it easy for a group to view.

Solarscope's optics are pretty clever. During

A group of people using the Solarscope can view the Sun's projected image. MICHAEL E. BAKICH

Daystar's Solar Scout SS60-ds is a Hydrogen-alpha telescope that will reveal prominences and details in the Sun's chromosphere. *ASTRONOMY* MAGAZINE

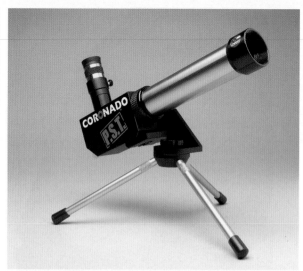

Meade Instruments' Personal Solar Telescope (PST) is ready to attach to a camera tripod right out of its case. The MALTA tabletop mount, shown here, is optional. *ASTRONOMY* MAGAZINE

assembly, you insert a mirror into an aluminum holding device. This snaps into the base of the Solarscope. Because of the mirror's shape, the Sun's image is offset inside the box. The other optic is a simple "telescope" made of a plastic tube with a front lens but without an eyepiece. The telescope attaches to the hood. To focus the Sun's image, simply screw the telescope in or out.

It takes about 10 minutes to assemble the Solarscope, and using it is easy. Just set it on a table or a stand. You can place it on the ground, but raising it up makes the Sun easier for people to see. To point it, move the hood with the telescope up or down while turning the base side to side until the telescope points at the Sun.

Solarscope's Individual model projects an image of the Sun 3.2 inches (80 mm) in diameter, enough to show medium-size sunspots. The Education model projects one 4 inches (100 mm) across.

As Earth rotates and the Sun "moves," simple adjustments toward the west and either up or down are all that's required. And if teachers or group leaders want to use multiple Solarscopes, users can operate them without fear of eye damage.

DAYSTAR SOLAR SCOUT

If you want to see more than just sunspots on eclipse day, you'll need to move into the realm of Hydrogen-alpha (Hα) observing. Daystar Filters, a company that has been making such products for decades, has introduced a new player—the Solar Scout SS60-ds—into that arena. This 2.4-inch (60 mm) dedicated solar scope is an all-in-one unit and

The black ring allows users of Coronado's PST to "tune" the image. A gentle turn reveals solar prominences; turn the ring the other way, and you'll see chromosphere detail. *ASTRONOMY MAGAZINE*

the ultimate in portability. You can purchase just the scope or a bundle, which includes a mounting foot, a star diagonal, an eyepiece, a battery, and more.

This scope sports a focal length of 930mm, which allows for high-quality, medium-power views of the Sun. For example, a 9mm eyepiece will yield a magnification of 103×. The SS60-ds also features a helical focuser that can't slip and a "Solar Bullet" finder, which makes centering the Sun's image in the scope easy. Power comes from a 110-240VAC plug-in adapter.

CORONADO PST

Another low-price self-contained Hα solar telescope is Meade Instruments Personal Solar Telescope (PST). The PST is a 1.6-inch refractor that comes

Meade Instruments' SolarMax II 60 is a Hydrogen-alpha telescope. Unlike a visible-light solar filter, which can show sunspots, the SolarMax II's filter reveals prominences, flares, and the Sun's chromosphere. ASTRONOMY MAGAZINE

with an 18mm Plössl eyepiece. It has a small knob to control focusing and a ring at the base of the brass tube that allows the user to tune the image. Tuning allows for better views of certain parts of the solar disk. Tune the PST's filter slightly, and the chromosphere is more prominent; tune it a bit differently, and prominences will stand out.

There is also a solar finder, which the maker calls a "Sol Ranger," built into the telescope's body. When you point the PST in the general direction of the Sun, a small image of the Sun appears in a circular window of semitransparent material near the eyepiece—a totally safe view. When you center that image, the Sun is in the telescope's field of view. The PST offers crisp, clear images with no sign of ghosting (a faint copy or copies of the main image) and no significant scattered light in the field of view. The Sun appears bright, but not so bright that features looked washed out.

MEADE SOLARMAX II 60

The SolarMax II 60 Telescope is an instrument made by Meade. This is a pricier option than the PST, so I'm going to leave in more of the tech speak because only advanced amateur astronomers will probably consider this option.

This refractor features 2.4 inches (60 mm) of aperture, a focal length of 400mm, and a focal ratio of f/6.7. Its filter is a two-piece, full-aperture etalon—an optical interferometer that bounces light between two partially reflective mirrors.

The lever to the right controls the tunable etalon in the SolarMax II. It tilts the front filter stack and thus moves the center of the 0.5-angström transmission band. One setting will provide a good view of the Sun's prominences, while another will show surface details better. ASTRONOMY MAGAZINE

When the telescope is not in use, a threaded metal cover protects the objective and the etalon. The SolarMax II system includes a blocking (energy-rejection) filter, the 0.5Å etalon, a diagonal, a 25mm eyepiece, mounting rings, a Sol Ranger solar finder scope, and a carry case. A dovetail mounting plate connected to two rings makes it simple to attach the scope to a mount.

This scope makes surface details easy to see, including prominences along the Sun's edge. Feel free to substitute your eyepieces instead of the one Meade supplies.

Solar viewing through a "white-light" filter gives an observer a peek at the Sun's majesty. Pick one of these three Hα scopes, and you might find yourself at the eyepiece for hours on every clear day.

A proper solar filter is a must for shooting the Sun at any time except during totality.
GETTY IMAGES

25 TIPS FOR PHOTOGRAPHING THE ECLIPSE

During the last "Great American Eclipse," on August 21, 2017, hundreds of thousands of people photographed the event. Most got disappointing results. Here are some tips you can follow that might help you take a picture you'll be proud of.

1 • SELECT YOUR EQUIPMENT WELL IN ADVANCE.

You need to decide soon what you'll be shooting through. It could be a removable camera lens, a non-removable lens on a point-and-shoot camera, or the tube of a telescope. For telescope recommendations, see Chapter 9. For camera recommendations, please head to Chapter 10. You also must decide what you'll be shooting with. This boils down to three choices: a digital single-lens reflex camera (DSLR), a point-and-shoot camera, or your cellphone. Most importantly, you'll need an approved solar filter for your optics.

2 • PRACTICE WITH YOUR SELECTIONS WELL IN ADVANCE

Planning on taking your first pictures of the Sun on eclipse day is no plan at all. I'm not suggesting you head out every day from now until April 8, 2024, and photograph the Sun. But practice will pay off. Believe

A visit to a shop that sells telescopes will reveal lots of options. The sales staff at such stores are generally knowledgeable and can help you make the right choice. HOLLEY Y. BAKICH

me, eclipse day will have enough tension and drama without you adding to it by making mistakes.

While you don't have to drill your photo skills until you're a pro, you do need to at least reach the amateur level to have success on eclipse day. How do you do that? Simply go outside on a sunny day, aim your filtered camera at the Sun, and shoot. During your practice run, take a range of different exposures. It doesn't cost anything to shoot 100 images, so grab a shot for all your settings. You don't even have to keep track of them because today's cameras record all relevant details about each image. They store this information in an EXIF file. Lots of web photo apps like Flickr show this data, and you also can see it if you use Photoshop.

After shooting, examine all the images and narrow them down to the best two or three. That's when to look at the EXIF file. In addition to the date and time of the exposure, you'll find the aperture (f-stop), focal length, ISO setting, exposure time, and more.

If you're shooting in manual mode, you really won't have to worry about exposure times until the eclipse nears totality. The Sun's surface brightness remains pretty constant throughout the eclipse, so you won't have to change exposure times until the Sun's disk narrows into a thin crescent. At that point, add two more stops.

3 • DECIDE WHAT YOU WANT TO SHOOT

Page through the images in this book. Look through back issues of *Astronomy* magazine, or check out archived "Picture of the Day" images on the Astronomy.com website. See how previous photographers have captured total solar eclipses and decide if you want to try to replicate one of their shots.

Some imagers mount their cameras on a tripod and take single wide-angle shots during totality. These capture the Sun's corona and part of the earthly scene below.

Others take close-up shots of the Sun, either of the entire eclipse with all the partial phases or just a few minutes before and after totality. Most top-notch eclipse photographers who go this route will mount their cameras on motorized telescope mounts, which track the Sun's position as it moves across the sky.

Still others take a sequence of wide-angle shots that show the progression of the eclipse. For this, they aim the camera at the Sun's position at mid-eclipse, lock it into place, and shoot at equal intervals such as once every 5 minutes. Later, these shots can be arranged into a montage or combined into one picture with software like Photoshop.

4 • ARRIVE AND SET UP HOURS BEFORE THE ECLIPSE STARTS

To make the most of eclipse day, you can't be scrambling to set up in the moments before the big event. Instead, stake out a location well ahead. The advantages to setting up in advance are numerous. You can pick out a prime spot. You can set up your equipment, test it, and deal with any problems.

Plus, the extra time at the site will allow you to interact with other eclipse-watchers before you lose yourself in photography. You'll even have time for a pre-event meal.

5 • PHOTOGRAPH EVERYTHING

If you have access to a second camera, you might want to chronicle what's happening around you as

Eclipse sequences make some of the most spectacular images of the event. Most imagers shoot individual frames and later combine them with processing software like Photoshop.
DANIELE57C/WIKIMEDIA COMMONS

Well before the 2010 total solar eclipse, residents of Easter Island set up to sell food and memorabilia to eager eclipse watchers. Pictures like this help capture what was happening on Earth. HOLLEY Y. BAKICH

the event approaches. These photos won't interfere with your experience of the eclipse and can document the activities of your astronomy club, friends, or family.

6 • BRING EXTRA BATTERIES

Put extra batteries on your checklist now to avoid a shutdown of your equipment at a crucial moment. For every piece of equipment that takes batteries, pack extras to have on hand—and make sure any devices that need to be charged have been fully charged in advance.

7 • CHECK YOUR CAMERA'S MEMORY CHIP

On a similar note, be absolutely sure that the memory card in your camera has at least twice the amount of free memory you think you'll need to adequately photograph the eclipse. In fact, why not purchase an additional memory card with at least 32 gigabytes of memory. The cost for a new 32Gb

A high-capacity memory card like this one won't cost much. Pick up a couple before the eclipse, and you'll have all the storage you need. MICHAEL E. BAKICH

card as I write this is $10, and $21 for a 64Gb card.

8 • PREP DURING FIRST CONTACT

The entire eclipse lasts more than 2 hours 19 minutes at all locations throughout the U.S., but almost half of that time is a slow build-up. Remember, this event is all about totality. If you're concentrating on capturing those 3 to 4 minutes of darkness, the hour after first contact is your final check to see that everything is working. You can still be social, but rehearse your plan at least a few more times. Then, about 25 minutes before totality, every great eclipse photographer I know moves into "image mode."

9 • SECURE YOUR FILTER

Most solar filters made for telescopes fit tightly on the front of the tube. Camera lenses, however, come in a wider variety of sizes. If your filter is even slightly loose, secure it with blue painter's tape. That product, available at any home improvement store, will not

It's all about totality.

10 • TOTALITY LASTS 8 SECONDS

Not really. But I often quote an old friend of mine, Norm Sperling, who wrote a "Forum" in the August 1980 issue of *Astronomy* magazine titled, "Sperling's Eight-Second Law." I'll reproduce the beginning here.

"Everyone who sees a total solar eclipse remembers it forever. It overwhelms the senses, and the soul as well—the curdling doom of the onrushing umbra, the otherworldly pink prominences, and the ethereal pearly corona. And incredibly soon, totality terminates.

"Then it hits you: 'It was supposed to last a few minutes—but that couldn't have been true. It only seemed to last eight seconds!'"

Get the point? If you encounter a problem—any problem—that takes more than a few seconds, stop! Forget about photographing the eclipse, and gaze in wonder at what's going on above you.

11. CALCULATE THE FIELD OF VIEW FOR YOUR SYSTEM

So many DSLRs exist today, and they all seem to have a variety of lenses they can accept. Plus, not all of them have the same size sensors (chips), and sensor size factors mightily into the field of view (FOV) calculation. So, instead of an unwieldy table, here's an easy way to figure out a lens' field of view on your camera.

FOV = 2 * arctan (0.5 * s/f) * 180/π
s = Sensor dimension in millimeters
f = Focal length of the lens in millimeters
180/π = 57.3 (close enough)

Do note that this formula is for just one dimension (width or height) of your chip. If you want the FOV in both dimensions, do this calculation twice. Here's an example.

Your camera is a Canon 6D with a 200mm lens. Camera buffs call the 6D's sensor (chip) either "full-frame" or "35mm." That means it measures the same as a piece of film in an old 35mm camera, or 36 by 24 millimeters.

The horizontal FOV would be 2 * arctan (0.5 * 36/200) * 57.3 = 10.3°. Now let's do the vertical FOV. While you could use the formula again, why not make things easy? The vertical measures 24/36 = 0.6667 of the horizontal, or 10.3° * 0.6667 = 6.9°. Therefore, the FOV for just this lens attached to just this camera measures 10.3° by 6.9°.

200mm 135mm*   **400mm 270mm***   **500mm 330mm***

1,000mm 670mm*   **1,500mm 1,000mm***   **2,000mm 1,500mm***

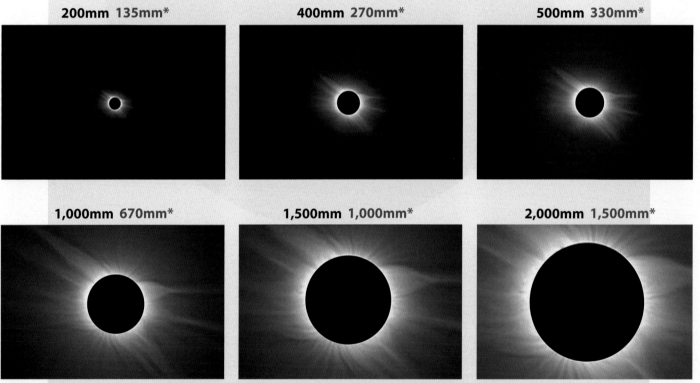

*Approximate focal length for most crop sensor DSLR cameras.

The correct focal length will help you capture the eclipsed Sun in as much or as little detail as you'd like. These simple figures show the image sizes of the eclipsed Sun for various focal lengths, as photographed with both a full-frame and a crop sensor DSLR. *ASTRONOMY* MAGAZINE/ROEN KELLY

leave a residue on your equipment. Securing your filter is vital, but so is the ability to easily remove it. One would-be photographer during the November 3, 1994, total solar eclipse had such a problem removing the filter from his camera lens that he ruined his careful alignment and was never able to return the eclipsed Sun to the field of view.

12 • A TELESCOPE'S FIELD OF VIEW

But what if you'll be attaching your Canon 6D body from Tip 11 to a telescope rather than to a camera lens? The calculation is the same. In this case, however, your telescope substitutes for a camera lens. So, you can still use the formula at left; just insert your scope's focal length. Most manufacturers print the focal length either on the tube or on the ring that secures the front optic. Not in either place? Then look in your instruction manual.

13 • GET THE RIGHT CAMERA-TO-SCOPE ADAPTER

This relates to the previous item. Technically, you'll be coupling your camera's body to the telescope's focuser, but you can't do that without a two-part adapter. You'll need both a T-ring and a T adapter. The T-ring, like the fitting on your lenses, has the male extensions that fit into your camera body's female slots. Turn it, and it will lock in place. Release it the same way you release a lens.

The T adapter screws into the T-ring (most celestial photographers just leave the two parts connected). The adapter's other side is a 1¼"-diameter tube that slides into your telescope's focuser just like a 1¼" eyepiece.

A T-ring (left) attaches to your camera like a lens does. A T-adapter screws into the T-ring. The smaller diameter chromed barrel goes into the telescope's focuser, taking the place of the eyepiece. MICHAEL E. BAKICH

14 • CALCULATE THE SUN'S SIZE ON YOUR CHIP

After you figure out the FOV of your camera/lens combination, it's a simple matter to calculate how much of your camera's sensor the Sun will cover. In the example in no. 11 at left, the width of the FOV is 10.3°. The Sun and Moon both have an angular diameter of 0.5°. So, the Sun's width on your image will be 0.5/10.3 = 4.9 percent (call it 5 percent) of the FOV.

15 • USE A REMOTE SHUTTER RELEASE

Why take the chance that you touching your camera will move it in a way you don't intend? Devices that trip your camera's shutter are small, easy to use, wireless, and inexpensive.

Eclipse photography becomes much easier and hands-off if you use a timer remote controller. MIKE D. REYNOLDS

16 • CALCULATE THE CORONA'S SIZE ON YOUR CHIP

If you're photographing totality, the Sun's disk is just the beginning. In fact, between second and third contacts you won't see its disk at all. What will still be visible, however, is the evanescent corona. It stretches between two and three solar radii from the Sun's edge. If we stick with the 6D and a 200mm lens, a corona measuring two solar radii will have a diameter of 2.5°. So, 2.5/10.3 = 24 percent, or one-fourth as wide as the FOV (and 36 percent as high). A three-solar-radii corona will span 34 percent, or one-third of the width and 51 percent of the height.

17 • DON'T GO NUTS ON YOUR LENS' FOCAL LENGTH

This item highlights the two points that precede it. If you want to capture the entire Sun to document the partial phases, but you are unable to make it to the path of totality and still want to photograph the partial eclipse above your site, make sure you don't pick a lens or telescope that restricts the view to less than 0.5° in the vertical dimension. Increase this to 3°, or even a bit more, if you want to photograph most of the corona. Remember, your camera will capture more than your eye can discern.

18 • FOCUS IS CRITICAL

Show someone an out-of-focus image you took of the eclipse, and they may not say anything about it. But I guarantee you they'll think, "Oh, wow, you spent all that time and money getting ready, and this is what you have to show for it?"

Today's camera lenses have the ability to focus past infinity, so you have to line up an arrow with the "sideways 8" on the lens. The easiest way to focus is to aim at a distant earthly object. Then don't touch the focus ring again. If this instruction makes you paranoid, you can tape the focuser using the same painter's tape I mentioned in no. 9. Also, turn off your lens' autofocus. Pointing it at the sky wreaks havoc with the high-tech focusers manufacturers use today.

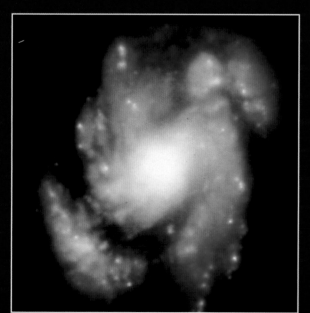 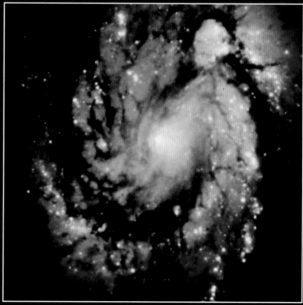

If you don't think focus is important, compare these images of spiral galaxy M100 captured by the Hubble Space Telescope. The one on the right was taken after astronauts installed a new optical component. NASA

19 • AVOID SMALL F-RATIOS

If you're using a standard zoom lens or have added a teleconverter to your system, the image will not be sharp if you choose to shoot at f/1.4. Instead, pick an f/ratio from f/8 to f/11, and you'll get the sharpest images you can. Fortunately, we're dealing with the Sun, so there's plenty of light to go around even through such reduced apertures.

20 • GET ACQUAINTED WITH YOUR WRITE TIME

How fast you can take pictures depends on your camera-to-memory-card write time. The latest cameras have microprocessors and internal memories that have reduced this interval to next to nothing. Older digital models and many point-and-shoot cameras, however, are another story. Be familiar with how your camera performs lest it delays your shots and surprises you by reducing the number of images you can capture.

21 • CONSIDER AN INTERVALOMETER

An intervalometer is a device that lets you take time-lapse photos. If your plan is to capture close-up images during the entire eclipse, an intervalometer will let you take exposures in any interval you choose. Search online, and you'll find a variety of intervalometers, also called timer remote controls, for many Canon and Nikon DSLRs.

Michael constructed this safe solar filter by attaching Baader AstroSolar Safety Film to a hardwood frame. The unit fits snugly over the front end of a telescope, and the black knob (attached to a threaded rod) secures it.
MICHAEL E. BAKICH

22 • USE AN APPROVED SOLAR FILTER

As should be clear from how often it has been repeated, viewing and photographing the eclipse must only be done through an approved solar filter. For your eyes to be safe and comfortable visually, a filter must reduce the visible light by a factor of 160,000. Camera chips are a bit more forgiving, and the unit's ability to adjust exposure times pretty much guarantees success. So, attach a filter to your camera lens.

23 • NO FLASH

If you're using a point-and-shoot camera, be sure to turn off the flash. Leaving it on not only won't do any good, but it will also drain the camera's battery and may even annoy the people near you.

24 • TAKE THE FILTER OFF TO CAPTURE THE CORONA

The Sun's disk outshines the corona by 1 million to 1. That's why we never see the corona except during totality. To photograph the corona, you must remove the filter from your lens (or telescope).

One thing to keep in mind when you're thinking about capturing the corona is that its brightness varies according to its distance from the Sun. It's really bright next to the edge, but it gets progressively fainter as you move farther away. Usually, by the time the distance is about one solar diameter, human eyes can no longer see the corona, but cameras can. So, take the time to bracket your exposures during totality. That way some will reveal detail in the inner corona while others show features in the much less dense outer corona.

25 • USE A STURDY TRIPOD

No eclipse photography falls under the "handheld" variety. Shoot away with your cellphones and point-and-shoot cameras if you want, but you're wasting your time. While image stabilization sounds good and is incorporated into upscale lenses, nobody experienced trusts it for capturing eclipses.

You have two choices for eclipse photography. First, you can use a standard tripod. Or you can attach your camera to a telescope mount (with or without the telescope) that sits on a tripod.

AND HERE'S A BONUS

PROCESS AND SHARE YOUR IMAGES QUICKLY

Can you imagine how many people will be photographing the eclipse? If you want your images to stand out or even to be published somewhere, whether in *Astronomy* or elsewhere, you have to send them out soon after the eclipse is over, certainly no later than the end of the day April 8. Email the best to ReaderGallery@Astronomy.com.

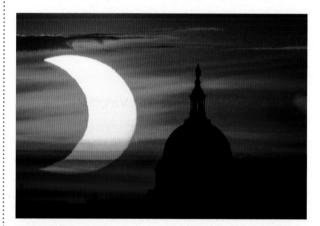

The U.S. Capitol building in Washington, D.C. adds a counterpoint to a partial eclipse in 2021. BILL INGALLS/ NASA

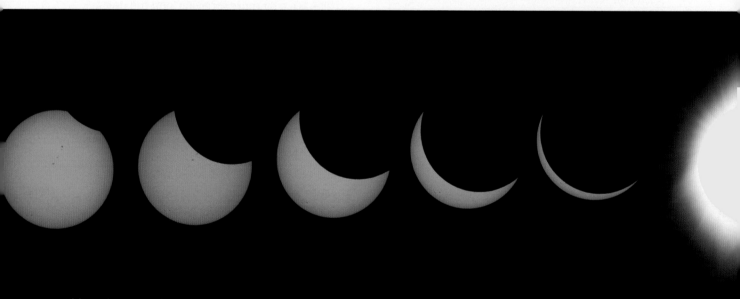

This series was taken above Madras, Oregon, during the August 21, 2017, total eclipse. AUBREY GEMIGNANI/NASA

When you photograph the partial phases, attach a solar filter to the front of your telescope. But make sure to remove it to shoot totality.

MICHAEL E. BAKICH

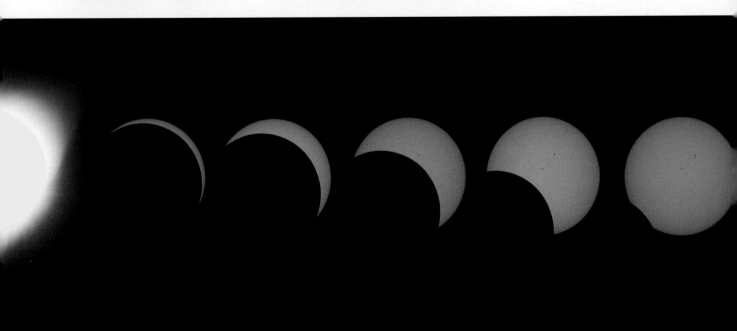

Rio Nazas (the Nazas River) provides a picturesque setting near Nazas, Mexico, where the 2024 eclipse will have its longest duration of totality.

KUNSTPHOTO/GETTY IMAGES

20 ECLIPSE HOT SPOTS

The total solar eclipse set to occur on April 8, 2024, will dazzle everyone who views it. However, potential observers might have some questions. Where exactly in Mexico and the U.S. will totality be visible? That's easy to answer with a detailed map. But which locations are the best spots to view the event? That answer is less straightforward.

You'll want to set up shop near the center line of the eclipse, where totality will last the longest. But what else makes for a good viewing site for a total solar eclipse? Here are 20 great locations you should consider for the 2024 Great North American Eclipse, starting in Mexico and working toward the Northeast U.S.

I'll also offer one other suggestion when choosing a site: Carefully consider the population. All things being equal, a town of 10,000 is much more likely to have event-related problems than a city of 75,000. Traffic will be one of the primary issues; small communities with one main road may suffer hours of gridlock. If you opt to travel to such a location, get there early—perhaps even a day or two ahead of the eclipse. (Remember: April 8, 2024, is a Monday, and most people will be free the entire weekend beforehand.) Drive safely and don't forget your eclipse glasses.

1 • MAZATLÁN, SINALOA, MEXICO

The Moon's umbra touches the coast of the United Mexican States at 12:07 p.m. Mexican Pacific Daylight Time less than 12 miles (19 km) southeast of Mazatlán, which was one of the main destinations for travelers viewing the July 11, 1991, total solar eclipse. This city knows how to host a large influx of travelers, so it's a good bet that it will be a prime destination.

Eclipse starts: 9:51:28 a.m. MPDT
Eclipse ends: 12:32:11 p.m. MPDT
Maximum eclipse: 11:09:39 p.m. MPDT
Sun's altitude at maximum eclipse: 69.1°
Duration of totality: 4 minutes 18 seconds
Width of Moon's shadow: 123.7 miles
(199.1 km)

2 • NAZAS, DURANGO, MEXICO

At 1:15 p.m. Mexican Central Daylight Time, the umbra arrives at Nazas. This town of some 4,000

Stand outside the Capt. Charles Schreiner Mansion in Kerrville, Texas, on eclipse day, and you'll enjoy 4 minutes and 24 seconds of totality. LARRY D. MOORE/WIKIMEDIA COMMONS

residents will surely see that number swell because it is the nearest location to the point of longest totality. The town itself will enjoy that duration, but the actual spot is about 3 miles (5 km) to the north, just east of Durango Paso Nacional, the road that connects Nazas to San Luis del Cordero. If you're headed here for the longest possible totality, get there at least a day early.

Eclipse starts: 11:58:24 a.m. MCDT
Eclipse end: 2:39:42 p.m. MCDT
Maximum eclipse: 1:17:17 p.m. MCDT
Sun's altitude at maximum eclipse: 69.8°
Duration of totality: 4 minutes 28 seconds
Width of Moon's shadow: 122.6 miles (197.4 km)

3 • PIEDRAS NEGRAS, COAHUILA, MEXICO

Piedras Negras is a large city that lies across the Rio Grande River from Eagle Pass, Texas. Anyone from the U.S. who wants to experience the maximum duration of totality (4 minutes 28 seconds), will either cross the Eagle Pass–Piedras Negras Interna-tional Bridge or the Camino Real International Bridge. Using Piedras Negras as a base is a good idea because its metro population is a quarter million, so it contains lots of amenities for travelers.

Eclipse starts: 12:10:08 p.m. CDT
Eclipse end: 2:51:17 p.m. CDT
Maximum eclipse: 1:29:37 p.m. CDT
Sun's altitude at maximum eclipse: 68.6°
Duration of totality: 4 minutes 25 seconds
Width of Moon's shadow: 120.9 miles (194.5 km)

4 • RADAR BASE, TEXAS

For eclipse chasers who want the greatest possible amount of totality without leaving the U.S., consider Radar Base, which lies right on the U.S.–Mexico border. Its name, by the way, isn't that of a military base but instead a small town of several hundred residents. That number will balloon on eclipse day, so be sure to get there early.

Eclipse starts: 12:10:26 p.m. CDT
Eclipse end: 2:51:30 p.m. CDT
Maximum eclipse: 1:29:53 p.m. CDT

Sun's altitude at maximum eclipse: 68.5°
Duration of totality: 4 minutes 27 seconds
Width of Moon's shadow: 120.9 miles
(195.5 km)

5 • KERRVILLE, TEXAS

While San Antonio certainly will be the base of operations for many eclipse chasers, most will not stay there, opting instead to head to the center line for an additional minute of umbral darkness. Several small towns lie centered in the path, the largest of which is Kerrville, with roughly 24,000 residents. Be sure to check in advance for any eclipse-related activities.

> **Eclipse starts:** 12:14:43 p.m. CDT
> **Eclipse end:** 2:55:29 p.m. CDT
> **Maximum eclipse:** 1:34:17 p.m. CDT
> **Sun's altitude at maximum eclipse:** 67.5°
> **Duration of totality:** 4 minutes 25 seconds
> **Width of Moon's shadow:** 120.2 miles
> (193.4 km)

6 • LAMPASAS, TEXAS

Many travelers will choose to base in Austin, the state capital of Texas. It's a good choice because that city features lots of amenities, especially in the food and music realms. Unfortunately, it lies near the southern limit of the path of totality. So, a good choice on eclipse day would be to drive the 68 miles (109 km) north to Lampasas, which lies quite near the center line.

> **Eclipse starts:** 12:18:03 p.m. CDT
> **Eclipse ends:** 2:58:23 p.m. CDT
> **Maximum eclipse:** 1:37:35 p.m. CDT
> **Sun's altitude at maximum eclipse:** 66.5°
> **Duration of totality:** 4 minutes 24 seconds
> **Width of Moon's shadow:** 119.7 miles
> (192.6 km)

7 • HILLSBORO, TEXAS

Although not a huge city, Hillsboro is an easy destination, lying, as it does, on Interstate 35 where I-35E and I-35W split south of Dallas. It also lies right along the center line of totality, which will help maximize your time under the Moon's umbra.

> **Eclipse starts:** 12:21:23 a.m. CDT
> **Eclipse ends:** 3:01:16 a.m. CDT

Maximum eclipse: 2:40:53 a.m. CDT
Sun's altitude at maximum eclipse: 65.5°
Duration of totality: 4 minutes 23 seconds
Width of Moon's shadow: 119.2 miles
(191.8 km)

8 • SULPHUR SPRINGS, TEXAS

Another great location in the Lone Star State is Sulphur Springs. It lies along Interstate 30, so getting there won't be a problem. Although the city isn't huge (less than 20,000 residents), there are many open areas. One is Coleman Lake and Park, which offers 186 acres with trails and waterfalls. Cooper Lake State Park lies 15 miles (24 km) north. It has more than 2,500 acres of land and nearly 20,000 acres of lake. Observing from a boat would certainly be relaxing. And you'll only lose 2 seconds of totality compared to Sulphur Springs.

> **Eclipse starts:** 12:25:38 p.m. CDT
> **Eclipse ends:** 3:04:52 p.m. CDT
> **Maximum eclipse:** 1:45:04 p.m. CDT
> **Sun's altitude at maximum eclipse:** 64°
> **Duration of totality:** 4 minutes 21 seconds
> **Width of Moon's shadow:** 118.4 miles
> (190.6 km)

9 • RUSSELLVILLE, ARKANSAS

With a population near 30,000, Russellville has enough resources to host a moderate influx of visitors for the eclipse. Those eclipse chasers who prefer to observe the event outside the city could head for nearby Mount Nebo, a flat-topped moun-

If it's clear April 8, 2024, historic Russellville, Arkansas, will be packed with enthusiastic eclipse watchers. PHOTOLITHERLAND/WIKIMEDIA COMMONS

tain that rises 1,350 feet (410 meters) above the surrounding valley.

Eclipse starts: 12:33:08 p.m. CDT
Eclipse ends: 3:10:46 p.m. CDT
Maximum eclipse: 1:52:10 p.m. CDT
Sun's altitude at maximum eclipse: 49.0°
Duration of totality: 4 minutes 11 seconds
Width of Moon's shadow: 117.2 miles (188.6 km)

10 • CAPE GIRARDEAU, MISSOURI

The largest city in southeastern Missouri that will experience totality is Cape Girardeau, with its 80,000 residents. It lies on the bank of the Mississippi River and is easily accessible from Interstate 55. For an additional 4 seconds of totality, eclipseophiles can head 10 miles (16 km) west on State Route 72 to Jackson.

Eclipse starts: 12:41:51 p.m. CDT
Eclipse ends: 3:17:26 p.m. CDT
Maximum eclipse: 2:00:21 p.m. CDT
Sun's altitude at maximum eclipse: 57.3°
Duration of totality: 4 minutes 6 seconds
Width of Moon's shadow: 115.5 miles (185.9 km)

11 • VINCENNES, INDIANA

This small city of some 16,000 residents lies along U.S. Highway 50 (east-west) and U.S. Highway 150, which becomes U.S. Hwy. 41 (north-south). It's a quick hop from Interstate 69. More importantly, it sits squarely on the eclipse's center line, so it will probably be a popular destination for inhabitants of the region.

Eclipse starts: 1:46:59 p.m. EDT
Eclipse ends: 4:20:57 p.m. EDT
Maximum eclipse: 3:04:55 p.m. EDT
Sun's altitude at maximum eclipse: 54°
Duration of totality: 4 minutes, 5 seconds
Width of Moon's shadow: 114.5 miles (184.3 km)

12 • INDIANAPOLIS, INDIANA

The umbra will cover a wide swath of Indiana, but most of the attention will focus on the state's capital city. Downtown Indianapolis is a metropolis served by four interstate highways and will surely be one

Downtown Lima, Ohio, is large enough to host many visitors on eclipse day. NYTTEND/WIKIMEDIA COMMONS

of the most sought-after destinations. It offers plentiful lodging, excellent cuisine, and many attractions for travelers.

Eclipse starts: 1:50:31 p.m. EDT
Eclipse ends: 4:23:10 p.m. EDT
Maximum eclipse: 3:07:56 p.m. EDT
Sun's altitude at maximum eclipse: 53°
Duration of totality: 3 minutes 49 seconds
Width of Moon's shadow: 114 miles (183.4 km)

13 • LIMA, OHIO

This small city of 36,000 is well-positioned for viewing the eclipse and is just large enough to handle a moderate influx of visitors. For those who want the maximum possible length of totality, you'll get 6 additional seconds if you drive south on Interstate 75 to Wapakoneta, and an extra second if you continue south to the center line.

Eclipse starts: 1:54:51 p.m. EDT
Eclipse ends: 4:26:01 p.m. EDT
Maximum eclipse: 3:11:43 p.m. EDT
Sun's altitude at maximum eclipse: 50.8°
Duration of totality: 3 minutes 51 seconds
Width of Moon's shadow: 113 miles (181.9 km)

14 • CLEVELAND, OHIO

With a metro population of more than 2 million, this city will host a multitude of eclipse chasers. Get there a couple of days early and fill the waiting time with visits to some of Cleveland's highlights, including the Cleveland Museum of Arts and the Rock and Roll Hall of Fame.

Eclipse starts: 1:59:20 p.m. EDT

Eclipse ends: 4:28:57 p.m. EDT
Maximum eclipse: 3:15:37 p.m. EDT
Sun's altitude at maximum eclipse: 48.6°
Duration of totality: 3 minutes 49 seconds
Width of Moon's shadow: 111.9 miles
(180.1 km)

15 • ERIE, PENNSYLVANIA

The only large city in the Commonwealth of Pennsylvania to be graced by the Moon's umbra is Erie, which, with its 100,000 residents, sits on the shore of the Great Lake that bears its name. It's certain that many eclipse chasers from Pittsburgh, 130 miles to the south via Interstate 79, will visit for the event.

Eclipse starts: 2:02:23 p.m. EDT
Eclipse ends: 4:30:48 p.m. EDT
Maximum eclipse: 3:18:12 p.m. EDT
Sun's altitude at maximum eclipse: 47°
Duration of totality: 3 minutes 42 seconds
Width of Moon's shadow: 111.2 miles (179 km)

16 • NIAGARA FALLS, NEW YORK

If the Northeastern United States has good weather on eclipse day, the most picturesque images of the event might come from Niagara Falls. One of the best perspectives will come from the outlook called Terrapin Point, where the Sun will hang halfway up in the southwest—directly over the Falls! Science buffs who observe or photograph the eclipse from this area will surely want to visit the Nikola Tesla Monument within Queen Victoria Park on the Canadian side of Niagara Falls. It lies only 0.3 miles (0.5 km) north of Terrapin Point.

Eclipse starts: 2:04:50 p.m. EDT
Eclipse ends: 4:31:57 p.m. EDT
Maximum eclipse: 3:20:02 p.m. EDT
Sun's altitude at maximum eclipse: 45.6°
Duration of totality: 3 minutes 31 seconds
Width of Moon's shadow: 110.8 miles
(178.4 km)

Position yourself near the Mars Hill Wind Farm on eclipse day, and you'll experience 3 minutes 10 seconds of totality just before the Moon's umbra leaves the U.S. MICHAEL SURRAN/ WIKIMEDIA COMMONS

17 • BUFFALO, NEW YORK

The largest city in New York that will experience the Moon's umbra is Buffalo, with its metropolitan population of 1.1 million. The center line passes right through downtown, so expect all activity to come to a screeching halt in mid-afternoon. Travelers desiring information about the eclipse might want to check with the staff at Zygmunt Planetarium, which is part of the Buffalo Museum of Science.

Eclipse starts: 2:04:54 p.m. EDT
Eclipse ends: 4:32:07 p.m. EDT
Maximum eclipse: 3:20:11 p.m. EDT
Sun's altitude at maximum eclipse: 45.6°
Duration of totality: 3 minutes 45 seconds
Width of Moon's shadow: 110.7 miles (178.2 km)

18 • PLATTSBURGH, NEW YORK

This small city of roughly 20,000 residents makes this list primarily because it's a one-hour drive from Montréal, Canada's second-most populous city. Montréal itself will enjoy nearly 2 minutes of totality, but all serious eclipse chasers will head south to the center line for that additional 90 seconds. Good choice.

Eclipse starts: 2:14:02 p.m. EDT
Eclipse ends: 4:37:04 p.m. EDT
Maximum eclipse: 3:27:29 p.m. EDT
Sun's altitude at maximum eclipse: 40.4°
Duration of totality: 3 minutes 33 seconds
Width of Moon's shadow: 108.4 miles (174.5 km)

19 • SHERBROOKE, QUEBEC, CANADA

Those Canadians who may not wish to cross the border can opt for Sherbrooke, which is only a 100-mile (161 km) drive from Montréal. With a metro population of nearly a quarter-million, Sherbrooke offers plenty of lodging and other amenities. And a quick 10-mile (16 km) drive south will bring you to the center line and 5 additional seconds of totality.

Eclipse starts: 2:16:35 p.m. EDT
Eclipse ends: 4:38:13 p.m. EDT
Maximum eclipse: 3:29:23 p.m. EDT
Sun's altitude at maximum eclipse: 38.8°

Duration of totality: 3 minutes 25 seconds
Width of Moon's shadow: 107.8 miles (173.5 km)

20 • MARS HILL, MAINE

To be honest, Mars Hill is a small town of some 1,500 residents. But just think of it: an amateur astronomer watching the Moon cover the Sun from a place named Mars Hill? Terrific. This location also is one of

Buffalo, New York will be the largest city in the Empire State to experience totality.
ATOMAZUL/DREAMSTIME.COM

the last spots in the U.S. to see totality. But if you're one of those serious types, just drive 20 miles (32 km) south for an additional 10 seconds of totality.

Eclipse starts: 2:22:20 p.m. EDT
Eclipse ends: 4:40:52 p.m. EDT
Maximum eclipse: 3:33:41 p.m. EDT
Sun's altitude at maximum eclipse: 35.2°
Duration of totality: 3 minutes 12 seconds
Width of Moon's shadow: 106.2 miles (171 km)

A partial solar eclipse in Arlington, Virginia, in 2021 is partially obscured by clouds. While it can make a dramatic photo, cloudy weather isn't the best forecast for eclipse watchers. NASA/BILL INGALLS

WEATHER PREDICTIONS

The information in this chapter comes from the website of highly regarded Canadian eclipse meteorologist Jay Anderson (www.eclipso phile.com), which he allows anyone to reproduce. Anderson first produced a study of weather prospects for the solar eclipse on February 26, 1979, and he's been doing it ever since.

MEXICO

The Moon's shadow reaches the coast of Mexico at Isla Socorro, the first of a half-dozen small islands before the continent. The marine environment produces lots of clouds, but cloudiness declines rapidly as the eclipse track approaches the mainland beaches near Mazatlán.

The dry season of Northern Mexico in April promises lots of sunny weather for the eclipse. Centerline cloudiness barely budges from the 25 to 35 percent range from Mazatlán to the start of the Sierra Madre Oriental. Only when the track reaches the east side of the Orientals and descends onto the Gulf Plains does the cloud cover rise, from around 30 percent at Monclova to near 50 percent at the U.S. border.

Cold fronts from Texas and New Mexico are a big problem because they can bring large areas of clouds. Some bring clumpy mid- and high-level clouds. More active ones bring large cloud shields with showers and thundershowers.

On sunny but unstable days, clouds may form on the mountain ridges and on east-facing slopes. The largest of these storms may send clouds across the Mexican Plateau to plague eclipse seekers. Finding sunny weather can be challenging, but the road from Durango to Torreón, which offers a route that parallels the center line for 150 miles, might provide a means to escape into sunshine. Fortunately, thunderstorms are more likely to build after the

This map, right, shows the mean cloud cover in April along the eclipse path in Mexico. NASA

The graph below displays the average April cloud amount over locations along the center line in Mexico. NASA

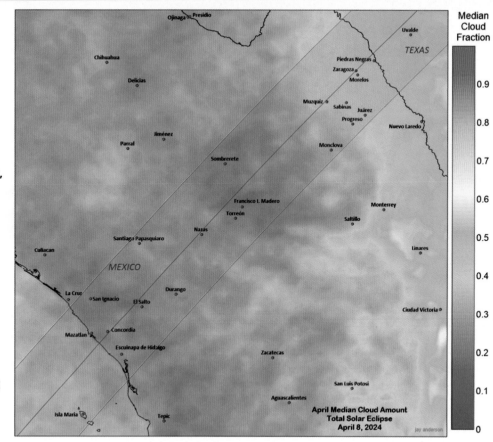

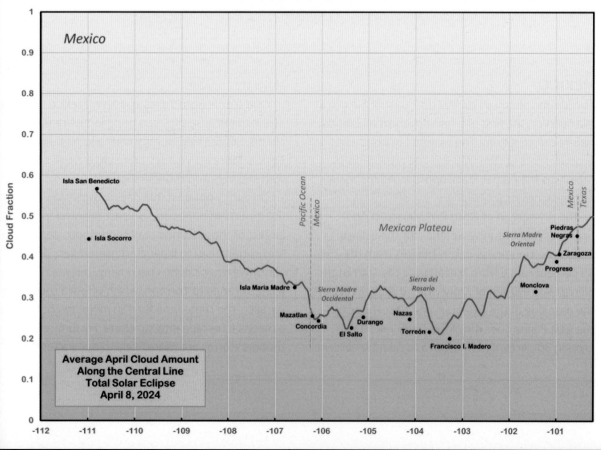

	Tmax (°C)	Tmin (°C)	Thigh (°C)	Tlow (°C)	Pcpn (mm)	Snowfall (cm)	Days with Pcpn	% Sunshine
Mexico								
Mazatlan	32	16	35	9	1.6		0.6	67
Durango	27	8	36	-6	6.2		1	78
Torreón	33	16	41	2	14		1.7	62
Monclova	32	17	42	4	23		2.7	58
Piedras Negras	33	15	42	-1	50		4.5	61

These statistics show Tmax, the average daily high; Tmin, the average low. Thigh and Tlow are the highest and lowest daily temps in the past 30 years; % sunshine is for the entire day. NASA

passage of the lunar shadow.

On the narrow coastal strip around Mazatlán, convective clouds are rare and southward-sweeping cold fronts make little impact. While the Mazatlán coast doesn't have the greatest sunshine along the track, it's only 3 to 5 percent cloudier than measured at Durango and Torreón.

The least promising eclipse-watching sites in Mexico come as the shadow approaches the Texas border. Low-level moisture from the Gulf of Mexico often moves northwestward toward the eclipse track beyond Nuevo Laredo. On most days, it

doesn't quite make it to the track, and as the day warms the cloud tends to retreat a bit. The presence of the coastal moisture and the greater exposure to cold fronts dropping southward from Texas is the main reason for the 15-percent growth in cloudiness as the eclipse track reaches the U.S. border.

TEXAS TO MISSOURI

Across Texas, the north side of the track has a notably sunnier April climate than the south. The best of Texas weather prospects—in fact, the best

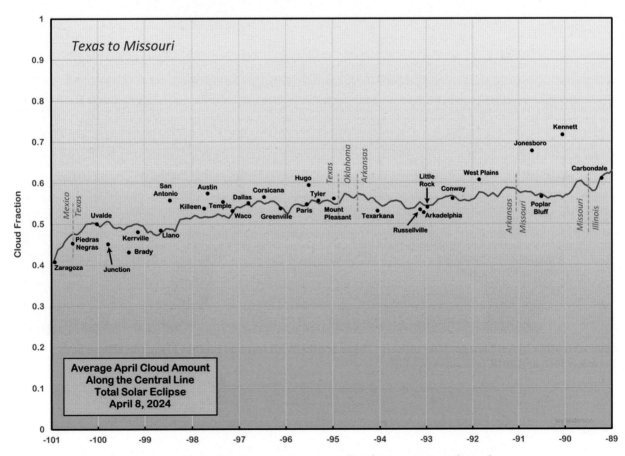

This graph shows the mean April cloudiness along the center line from Texas to Missouri. NASA

prospects in the U.S. and Canada—lies on the Edwards Plateau, where median cloud amounts are as much as 15 percent lower than those south of the center line on the Coastal Plain.

We can be even more specific: According to the satellite data, the best prospects lie between Junction and Brady, both in Texas. Brady is perilously close to the north limit, but Junction, 17 miles inside the track, has an eclipse duration of 3 minutes 7 seconds. Satellite measurements show a median April cloud fraction of 39 percent in the area.

Another good eclipse-viewing spot is one tucked up against the Mexican border where cloud amounts are only fractionally greater than at Junction, generally between 40 and 45 percent. The town of Eagle Pass, opposite Piedras Negras,

Mexico, which lies close to the umbral axis, may provide a convenient home base for those who wish to remain in the U.S. Another location near Llano is a low-cloud site for those who wish to immerse themselves right in the middle of the umbral shadow.

Past Llano, cloud amounts climb slowly along the shadow path, rising from 47 to 56 percent along the center line. Through Oklahoma, Arkansas, and most of Missouri, cloud cover along the track axis varies between 54 and 59 percent. It is not until the eclipse shadow is almost in Illinois that the cloud amount tops 60 percent.

The boundary between cloudy lowlands and sunnier heights is abrupt, especially in Missouri. Jonesboro and Kennett, which lie on the floodplain, have 10 to 15 percent higher April cloud amounts

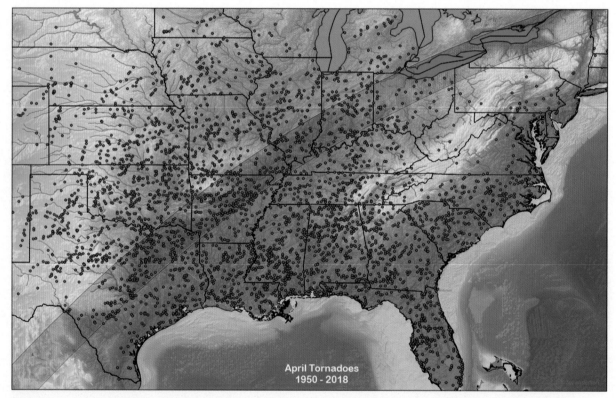

April Tornadoes
1950 - 2018

Severe weather in April is a genuine concern. These are the April tornado reports along the eclipse path between 1950 and 2018. STORM PREDICTION CENTER

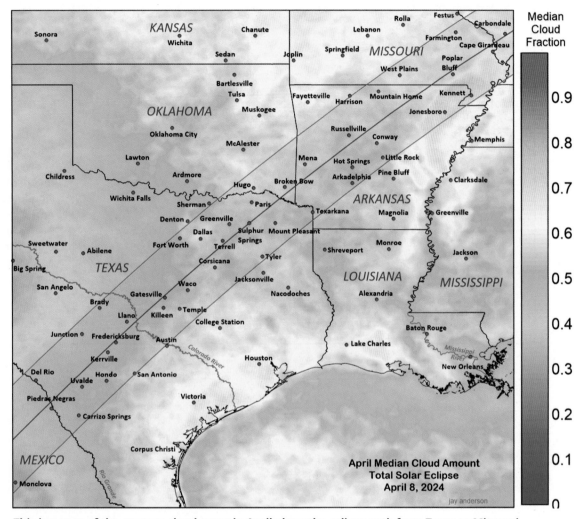

This is a map of the average cloud cover in April along the eclipse path from Texas to Missouri. NASA

	Tmax	Tmin	Thigh	Tlow	Pcpn	Snowfall	Days with Pcpn	% Sunshine
	(°F)	(°F)	(°F)	(°F)	(in)	(in)		
Texas								
Austin	80	59	99	30	2.09		7	54
San Antonio	81	57	101	31	2.6		7	56
Dallas-Fort Worth	77	51	92	21	3.4		7	61
Arkansas								
Little Rock	73	51	95	28	5.1		9	62
North Little Rock	74	53			4.3		11	68
Missouri								
St Louis	67	47	93	20	3.7	0.4	11	57

Here are the cloud statistics for selected locations along the eclipse path from Texas to Missouri. Tmax is the average daily high; Tmin, the average low; Thigh and Tlow are the highest and lowest daily temperatures in the past 30 years; % sunshine is for the entire day. NASA

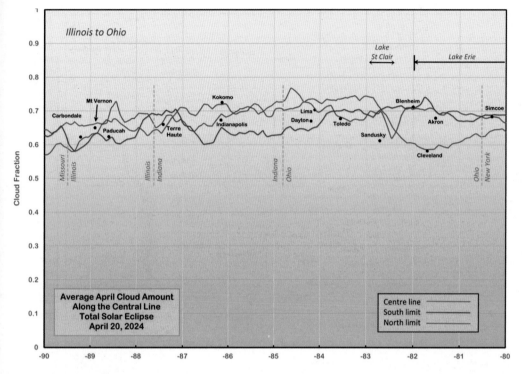

This graph shows the mean afternoon cloudiness in April along the center line and along the north and south limits. NASA

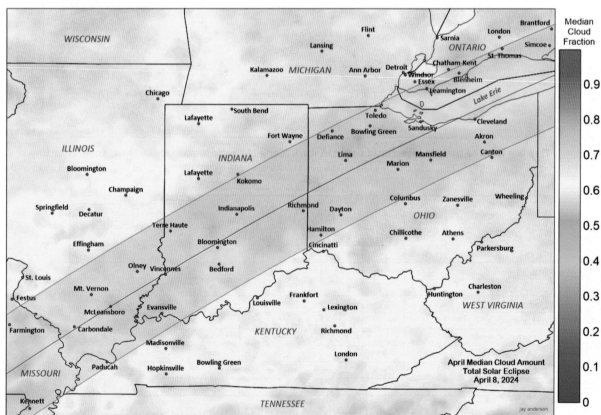

This is a map of the average cloud cover in April along the eclipse path from Illinois to Ohio. NASA

than stations on the north side of the track. From Little Rock northeastward, it's probably best to stay north of the center line and head for the Ozark Plateau unless the weather forecast promises a sunny day across the track. In Texas, about 7 days of the month report rain; in Arkansas and Missouri, rain comes on about 10 days.

ILLINOIS TO OHIO AND SOUTHWESTERN ONTARIO

As the eclipse track reaches Illinois, it moves into the path of the mid-latitude spring storms. At this latitude, they are a regular event, probably every three or four days. Along the portion of the track through Illinois, Indiana, Kentucky, and Ohio, there are no substantial elevation changes that would affect the cloud climatology. On the center line, cloud amounts vary between 60 and 70 percent from Illinois to Lake Erie, except for a modest 5-8

percent lower cloudiness along the south limit from Owensboro, Kentucky, to Cincinnati.

Just beyond Sandusky, Ohio, cloud percentage along the center line drops abruptly by more than 15 percent as the track reaches and crosses Lake Erie. The south-limit line, which passes 40 miles south of the lake, doesn't show this decline.

Unsettled days often see afternoon cumulus clouds and sometimes thunderstorms. Those clouds rely on heating from below, so the cold air over the lakes helps keep skies cloud-free over the lake and for a few miles inland.

The main beneficiary of this suppression of clouds are communities along the south and southwest shores of Lake Erie and south of Lake St. Clair in Canada. In particular, Cleveland lies close to the spot with the lowest median cloud cover. In Canada, communities between Leamington and Blenheim will profit from being downwind of Lake St. Clair.

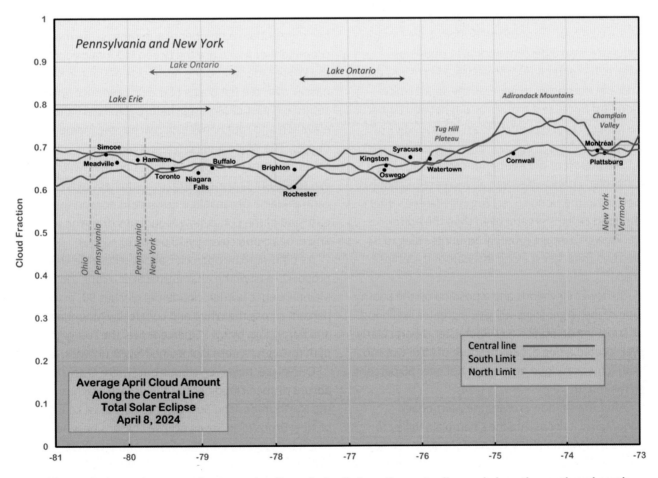

This graph shows the mean afternoon cloudiness in April along the center line and along the north and south limits. NASA

	Tmax	Tmin	Thigh	Tlow	Pcpn	Snowfall	Days with Pcpn	% Sunshine
	(°F)	(°F)	(°F)	(°F)	(in)	(in)		
Pennsylvania								
Erie	56	38	89	12	3.33	3.2	14	
New York								
Buffalo	55	37	94	5	3.01	2.7	13	48
Rochester	56	37	93	7	2.7	3.9	13	51
Syracuse	57	37	92	9	3.19	3.8	14	46
Watertown	58	37	93	13	3.17	2.3	12	
	(°C)	(°C)	(°C)	(°C)	mm	cm		
Ontario								
Toronto	12	4	32	-15	61	5	13	45
Hamilton	12	2	30	-13	71	8	13	45
Niagara	13	2	33	-10	66	6	12	41
Kingston	11	2	28	-13	71	7	13	40
Cornwall	12	2	31	-15	81	10	12	

Here's the climate data for selected locations along the eclipse track over Pennsylvania, New York, and Ontario. NASA

PENNSYLVANIA, ONTARIO, AND NEW YORK

Over Lake Erie, the center line shows reduced cloud amounts, but this tapers off toward the narrow end of the lake near Toronto and Buffalo. Lake Ontario's influence is an 8 percent decline in cloudiness where the center line approaches and crosses the shore at Rochester. In general, the south limit is cloudiest because it crosses a rougher and higher terrain and gets no benefit from the presence of the lakes.

The impact of the Adirondack Mountains is particularly noticeable, with an increase of about 10 percent in the cloud cover. Tug Hill Plateau's effect on cloud cover is greater even than the Adirondacks, with average monthly amounts rising to 84 percent, the highest along this part of the eclipse track.

So, lower elevations and a position on the south side of the larger lakes will give a greater likelihood of sunshine on eclipse day. Rochester is particularly well located along the south shore of Lake Ontario with a median cloud amount of just over 60 percent in the month.

After the shadow crosses the Adirondack Mountains, it drops into the Champlain Valley to move into Vermont. It's a 2,300-foot drop, causing the cloudiness along the center and south side of the shadow path to fall about 10 percent in response.

VERMONT, NEW HAMPSHIRE, QUEBEC, AND MAINE TO NEW BRUNSWICK

As the shadow path crosses into Vermont, cloud amounts climb. The south side of the eclipse path has to negotiate a number of sub-ranges, the most prominent of which are the Green Mountains and the White Mountains. The north side of the track passes over the flat plains of the St. Lawrence Valley south and east of Montréal before gaining a little elevation and roughness as it reaches the Maine border. The White and Longfellow Mountains generate higher amounts of clouds on their western sides. But there is a drop in cloudiness over central Maine because the downward flow on the east side of the Longfellow Mountains dries the air and drops cloud cover by 10 percent.

It's difficult to be optimistic about eclipse-viewing prospects east of the Champlain Lakes. Along the Maine-Québec border, cloudiness reaches 90 percent across the White and Longfellow Mountains and barely falls below 75 percent over the rest of the path through eastern Maine and New Brunswick.

One place, however, stands out. There is an abrupt drop in cloudiness as the eclipse reaches the Northumberland Strait on the northeast coast of New Brunswick. So, an eclipse chaser's best bet might be a location on the shores of the Gulf of Saint Lawrence. The advantage gained isn't large—about 15 percent—but it is the lowest level of cloudiness east of Lake Ontario.

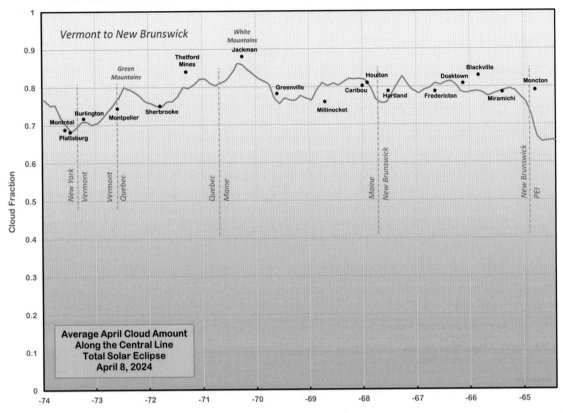

This graph shows the afternoon cloudiness in April along the center line from Vermont to New Brunswick. NASA

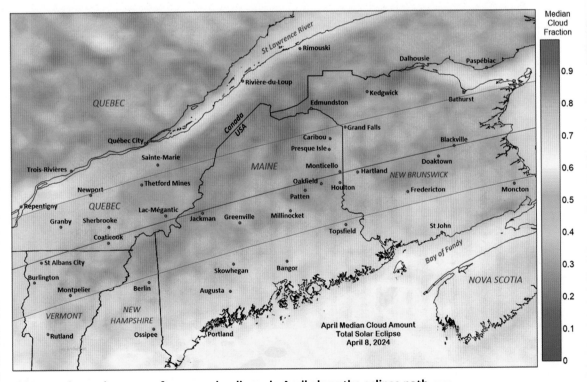

This map shows the mean afternoon cloudiness in April along the eclipse path. NASA

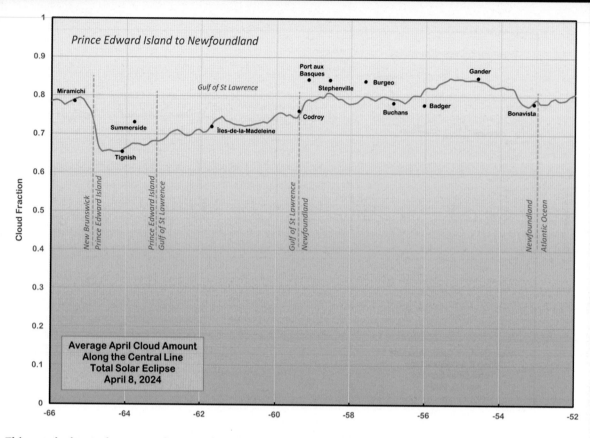

This graph shows the mean afternoon cloudiness in April from Prince Edward Island to Newfoundland. NASA

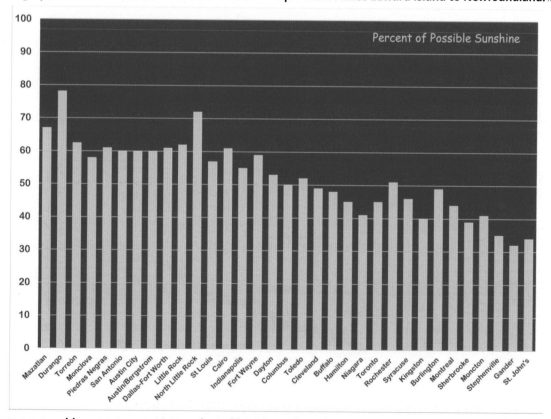

Here are sunshine measurements at selected locations along the eclipse path. NASA

THE GULF OF ST. LAWRENCE AND NEWFOUNDLAND

After crossing Northumberland Strait, the Moon's shadow passes Prince Edward Island (PEI) and heads across the Gulf of St. Lawrence. When the shadow reaches Newfoundland, the terrain rises quickly from the shoreline. The mountains are not particularly high, but they do influence the cloud cover. After passing the mountains, the shadow settles onto relatively flat terrain, moving across a sparsely treed landscape of rolling hills until it reaches Bonavista Bay and heads out across the Atlantic.

The north tip of PEI lies just south of the eclipse center line and this location, north of Tignish, is probably the best in Canada's Maritime Provinces from which to view the eclipse. The shoreline from North Cape to West Point benefits from the cold water of Northumberland Strait, with cloud amounts of just over 65 percent.

Beyond PEI, the cloud cover climbs to a discouraging 85 percent near Gander. Satellite images of Newfoundland show only one April 8 in the past 20 years where the whole island was clear.

ECLIPSE-DAY PLANNING

Seeing the April 8, 2024, total solar eclipse could be a challenge. This climate study is most useful for long-range planning but becomes less helpful as eclipse day approaches. By late March 2024, turn your attention to the regular long- and short-range forecasts available from a number of agencies.

Satellites maintain a continuous watch across North America and the satellite of choice for this event is GOES East, located over the equator and streaming images at 5-minute intervals. The website that does the most justice to the resolution of GOES East imagery is the College of DuPage site at weather.cod.edu. There you can look at an overview of the whole track, or you can zoom in to higher-resolution views of a particular state.

Satellite photos are OK for what's happening now, but they're limited for what will happen in the next day or two. For that information, we need computer forecasts. For single-point forecasts, one of the best is the Spotweather site at spotwx.com. Spotwx will give you a model forecast for a single location and will allow you to examine several different models to make some judgment about the reliability of a prediction.

When eclipse day arrives and the weather at your site is looking iffy, the local weather broadcast channel may be all that you need. For the most part, it concentrates on what's happening now and what will happen in the next few hours. In the end, preparation and mobility will be important. Good luck!

The lighthouse at Dalvay, in Cavendish National Park, on the north side of Prince Edward Island, Canada, might make a good spot to watch the eclipse. Prince Edward Island benefits from the cold waters of the Atlantic to reduce cloud cover compared to other sites in the region. © ONEPONY | DREAMSTIME.COM

Aqua satellite image of North America along the eclipse track April 8, 2017. NASA

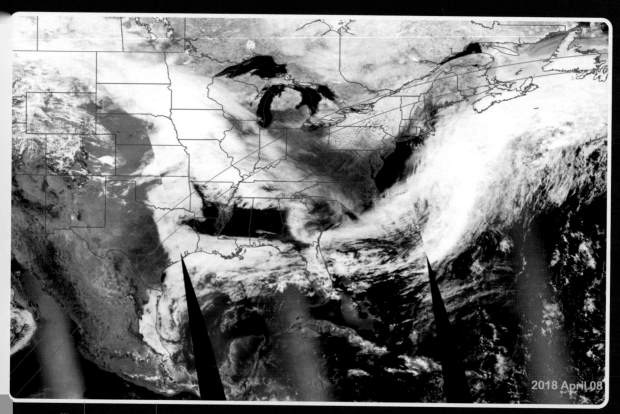

Aqua satellite image of Nort

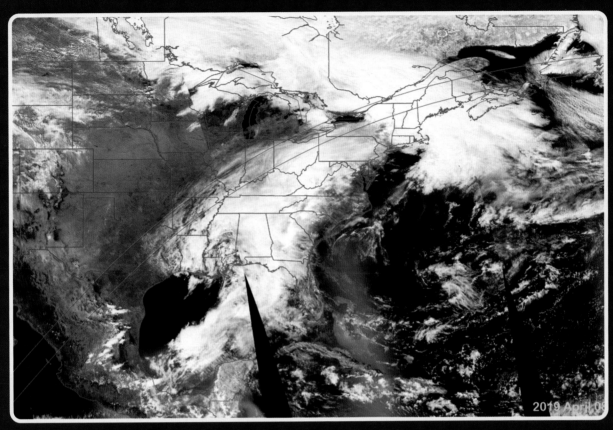

Aqua satellite image of North America along the eclipse track April 8, 2019. NASA

Aqua satellite image of North America along the eclipse track April 08, 2020. NASA

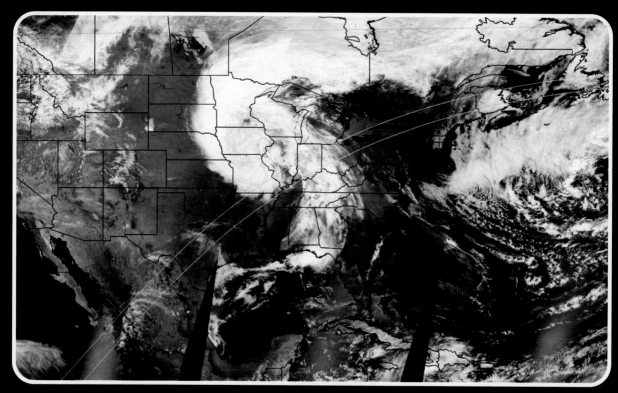
Aqua satellite image of North America along the eclipse track April 8, 2021. NASA

EXPERIENCE TOTALITY!

As I'm sure you know by now, the April 8, 2024, total solar eclipse is going to be a very big deal. Cities, counties, and even whole states are preparing for a massive influx of people from all over the world. Indeed, if the skies are clear along the path of the Moon's shadow, this could be the most observed total eclipse in history.

Don't get left behind. Don't be one of the people interested enough to read this book, but one who doesn't travel to totality. Because as I've said before, no partial eclipse—no matter how deep—can convey the drama and spectacle that happens when the Moon completely covers the Sun's face.

Remember, it's all about totality.

Composite images of solar eclipses can illustrate the event's progression and also capture earthly scenery. ALLAN WHETZEL

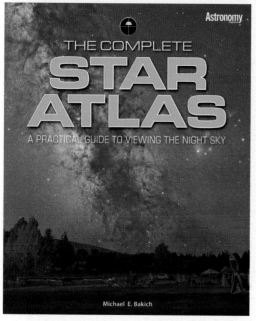
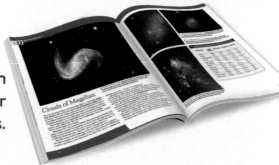